Preface

The idea for this book arose out of my work as a technical author with a company involved in, among other things, training technical personnel in writing and documentation. Commercial books on this increasingly important subject are remarkably thin on the ground. Hence the need for a volume such as this.

My thanks to R. C. DeVine and Nigel Biggs of Intereurope Technical Services at Wokingham for their help and advice on certain technical matters, and to John Field, Divisional Director, for offering complete access to the company's expertise.

Naturally, any errors or omissions which have eluded the net are entirely due to the shortcomings of the author.

My thanks are also due to Cambridge University Press for allowing me to quote from 'On the Art of Writing' by Sir Arthur Quiller-Couch, and to the British Standards Institution for permission to extract material from two British Standards.

Frensham, Surrey   J. E.

Beginner's Guide to
Technical Writing

John Evans

Newnes Technical Books

Newnes Technical Books

is an imprint of the Butterworth Group

which has principal offices in

London, Boston, Durban, Singapore, Sydney, Toronto, Wellington

First published 1983

British Library Cataloguing in Publication Data

Evans, John
  Beginner's guide to technical writing.
  1. Technical writing
  I. Title
  808'0666021   T11
  ISBN 0-408-01161-0

Photoset by Butterworths Litho Preparation Department
Printed in England by Butler and Tanner Ltd, Frome, Somerset

Contents

Introduction

The following entry appears in the Shorter Oxford Dictionary:

> *Author* 1. *gen*. The person who originates or gives existence to anything: a. An inventor, constructor, or founder. b. (*of all*, &c) The Creator (Middle English). c. He who gives rise to an action, event, circumstance, or state of things. d. The prompter or instigator – 1656.
> 2. *spec*. One who begets; a father, an ancestor (still used in *Author of his being*).
> 3. *esp*. and *absol*. One who sets forth written statements: the writer or composer of a treatise or book (now usu. includes *authoress*).
> 4. An authority, an informant.

A satisfactory definition of a *technical* author, however, requires a little more caution. A technical writer rarely instigates the work he performs – though there are exceptions. In the great majority of cases he is asked by a manufacturer or government department to produce the supporting documentation for an industrial or military product in its final pre-production stages. The department or organisation is the editorial and validating authority, and the author becomes part of the project team.

Occasionally, no doubt, he may chance upon a flaw in the design logic or an inconsistency in the maintenance philosophy, but such events are rare. A technical writer is often an

1

outsider, having been hired or recruited at the last moment. Nonetheless he has his own special part to play and to the end-user of the equipment he will be as important as the chief engineer.

Of the dictionary definition which began this Introduction, all that remains to a *technical* writer is paragraph 3:

One who sets forth written statements: the writer or composer of a treatise or book.

For accuracy, we may say:

One who sets forth written statements on technical subjects: the writer or composer of a technical treatise or book.

There are *some* technical writers who do originate their own material, and who fall into category 4:

An authority, an informant.

These are the writers of general technical books, educational textbooks, or articles for the scientific press. Our definition should be wide enough to include their work and that of all who engage in the writing of technical and scientific material.

A technical writer, then, may be said to fall within four definite categories:

- A writer working in the technical publications department of a manufacturing company or service ministry.

- A writer working for a specialist product support company.

- A freelance technical writer.

- A writer, occasional or otherwise, of technical books, articles, reports, or presentations, who might also be a manager, a student, an engineer, or even a housewife.

'Beginner's Guide to Technical Writing' was written with all these practitioners in mind, and not simply for those in the broad centre of technical writing.

Beyond the problems of definition, two other questions concern the prospective writer:

Firstly, it *is* a job, not a special calling. Efficiency always precedes inspiration.

Secondly, a writer may meet and deal with a wider mix of the population than many others in more apparently mobile occupations. A list of trades and professions a writer could have contact with might include:

- Management
- Designers
- Engineers
- Maintenance men
- Technical layout draughtsmen
- Tracers
- Printers
- Technical copy typists
- Document copyists and clerks
- Technical illustrators
- Photographers
- Editors

And last, but never least, that most agreeable of persons, the reader, who may include some, and occasionally all, of the above.

Technical and scientific communication is a very wide subject. So wide that is hardly a 'subject' at all. There are, of course, many ways to communicate. The approach adopted in this book is to paint with a broad brush, sketching in important details where necessary. As in most fields, competence comes from experience and a good technical background. The imponderables are left to the individual.

Outline

'Beginner's Guide to Technical Writing' develops the premise that a technical publication is like any other technical product. Consequently, the work involved in producing such a publication is divided into three phases:

- Design
- Development
- Production

Each phase is based on a flowchart outlining the steps used in its implementation. It is hoped that this may give some coherence to the rather eclectic amalgam of techniques generally involved in technical writing.

The book begins with a survey of the field: what constitutes technical writing, and the types of document that these categories contain.

The productive side of authorship, from the concept to the printing process, is considered under a three-phase division. Other matters related directly or indirectly to the work of a writer, such as technical illustration and some miscellaneous techniques (network planning, for example) are also discussed.

Each chapter ends with a summary of key concepts to supplement and add reinforcement to the text, and to aid subsequent revision. A glossary of important terms is given at the end, and those words which appear in the glossary are marked in the text with an asterisk. A section containing useful information – addresses, courses and so on – is added for completion.

Where a list is necessary as part of a section (e.g. specifications, standards etc.), this has been left in the text rather than shunted off to an appendix. The aim throughout has been to minimise footnotes and appendices and to provide a continuous text wherever possible.

1

Technical writing

The greater part of a technical writer's work is concerned with the writing and production of manuals and handbooks. These ubiquitous documents take many forms, from the explanatory leaflet in five languages to multi-volume sets in huge ring binders. They are vital to the efficient running of industry and the armed services. The first section of this chapter discusses manuals and handbooks, or, to give them a generic title, customer support documentation.

Reports, on the other hand, are often written at earlier stages in the production process. They may contain technical proposals or feasibility studies; progress reports on current developments; or post mortems on 'what went wrong'. Reports are usually generated internally, either by the technical publications (tech. pubs.) department of an organisation or by involved staff members. The second section covers the wide field of reports – or product documentation.

The third part of the chapter outlines the problems likely to be encountered by the writer of technical articles. Markets for this work, though fairly numerous nowadays, are neither lucrative nor easy to break into, and are best approached by the specialist journalist or very experienced amateur.

A field which sometimes escapes the net of more orthodox technical authorship is that of technical sales literature. Often presented in a glossy format and written in snappy language bristling with buzz words, these documents are mainly produced by copywriters employed by a bureau or advertising agency. Nevertheless, an occasional job of this type may fall into the technical author's lap, and a brief survey will not be out of place here.

There follows a consideration of the expanding field of technical training material – the 'programmed learning package', the training film and audio-visual*, and the educational textbook.

The final part of Chapter 1 concerns that most growing of growth areas, software documentation. 'Systems people' are much in demand nowadays and the art of software authorship is well worth careful study.

Manuals

Maintenance manuals and user guides, parts catalogues and operating handbooks are classified under the umbrella heading of *Customer Support Documentation*. Whatever the length and cost may be, they are written with a common aim, that of providing the customer with accurate and relevant data about a particular product.

The preparation of a technical manual involves a number of distinct steps. *Figure 1.1* gives an overall view of the procedures involved from initial requirement to printing. The series of event depicted here remains valid for even the simplest of authorship projects, and in later chapters it will be broken down into three essential phases: Design, Development, and Production. For the present we shall concern ourselves with the general format of a technical manual – what it looks like, and the specification governing it.

A specification* is a document prepared by an authority (Service Ministry or Standards Institution, for example) setting out in some depth the format*, presentation*, heading weights* scheme, style and shape of a technical handbook. Nothing – in theory – is left to chance, and the object is total standardisation between documents.

The standard British handbook specification is BS 4884* produced by the British Standards Institution. The Military equivalent, JSP (Joint Service Publications), are based on 4884 and will in time become the standard for all the services. Specifications are dealt with more fully in Chapter 2.

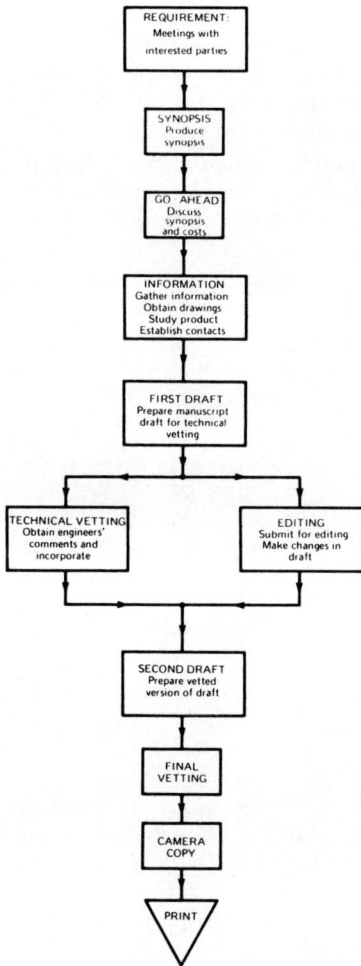

Figure 1.1 The preparation of a technical manual

In general, the terms of reference of any writing contract, including the specification, are laid down by the commissioning authority or customer. If this organisation is a commercial company, like Plessey or Ferranti, they may have their own in-house specs, in which case the author will be expected to conform to these. In the quite common event that the company is contracting for the Ministry of Defence (MOD), a military spec will be stated in the terms of the order.

The British Standards Institution defines the scope of BS 4884 as specifying 'information to be given in technical manuals and other documents designed to facilitate the use, maintenance and repair (where appropriate) of any material or product.' The Institution suggests a nine category division of the book, or books:

1	Purpose and planning information	(What it is for)
2	Operating information	(How to use it)
3	Technical description	(How it works)
4	Handling, installation, storage, transit	(How to prepare it for use)
5	Maintenance instructions	(How to keep it working)
6	Maintenance schedules	(What is done when)
7	Parts lists	(What it consists of)
8	Modification instructions	(How to change it)
9	Disposal instructions	(How to dispose of it)

These nine categories have been designed to cover all the information that a product's user is likely to need. The order of listing is to some extent arbitrary (though pertinent to some applications), and may be altered as necessary. The standard also explains that 'the extent of the information provided will depend on the nature of the product, the user's needs and the maintenance policy; some products will not require any supporting information and others will require only some of the categories.'

Manuals supporting highly complex equipment may consist of a single volume for each category. A simple appliance might use only one or two categories in a single folded sheet of paper.

As a structure for the organisation of a technical manual BS 4884 is a useful arrow in one's quiver. If a client has no specification in mind when approaching a writer, this system may be suggested. Both client and writer will then have a basis for further discussion, and expectations will be broadly in line with the material produced.

Further division of the nine categories into subdivisions turns BS 4884 into an extremely helpful pigeon-hole system in which to allocate the mass of technical data generated by modern industrial and military equipment:

1 **Purpose and planning information**
   Provides the user with information relating to suitability of the product for particular applications.
   Performance: Limits of performance. Handling require-ments, physical dimensions and weights. Limitations on environment, and reliability.
   Sources of information: References to documents, speci-fications, patents, and so on.
   Hazards: All known hazards.

2 **Operating information**
   Contains lucid instructions for operating the equipment under all conditions.
   General: Reasons for particular points in operating pro-cedure. Details of each user control. Manual/automatic/remote control. Safety precautions.
   Operating instructions: Step by step instructions for preparing, starting, operating and shutting-down the product in all modes under normal and emergency conditions.
   Hazards: Safety monitoring devices. Warnings of hazards arising from misuse.
   Malfunction: Procedures to be used during malfunc-tions, or when affected by external hazards.
   Disposal: Safe methods of disposal of part, or all, of the equipment, and the dangers involved.
   Correcting malfunction: Detecting malfunctions. Oper-ator correction procedures.

Planned sequential procedures (drills): Critical procedures. Strict drills for operation. Consequences of not following exact sequence.

Presentation: Procedures (drills) for a team of users – each member clearly delineated.

Practice drills: Exposition. Distinction between practice and operation.

3 Technical description

Provides descriptions on the functioning of 'state of the art'* equipment.

Purpose: General background and purpose of equipment.

Systems: Interface data for complex systems.

New techniques: Specific to unfamiliar designs. May be covered in an annex*.

4 Handling, installation, storage, transit

Contains the technical information required for storing or relocating the equipment.

Reception: Unpacking instructions. Special points.

Installation: Installing and setting up the equipment. Test schedules and performance parameters. Connection of services (electricity, etc). Precautions.

Setting to work: Preparation for use. Special tools. Test equipment.

Storage: Requirements for storage. Storage life. Tests and inspections. Use after transit.

Environment: Environment required for all uses.

Hazards: As considered necessary.

5 Maintenance instructions

Information based on likely resources available to the user. Special tools or materials. Maintenance procedures.

Tasks: Routine maintenance. Checks and inspections. Fault diagnosis and corrections. Overhaul.

Details: Instructions for each task. Warnings.

6 Maintenance schedules

Cycle of maintenance routines, and lists of tasks, time intervals or run-time of equipment.

Tasks: Schedules for each trade in order of frequency.

Complex products: schedules for a team of maintainers.

7 Parts lists

Contains information for location and identification of all replaceable items.

Content: Assemblies, subassemblies, replaceable parts. Illustrations of parts.

Identification: All relevant information allowing identification of part and/or re-ordering.

Other information: Availability of spares, maintenance levels of replacement.

Short lists: Lists of consumable, disposable, or short-life items, and kits for options or typical maintenance procedures.

8 Modification instructions

Modifications: Authorised changes and improvements. Separate publications may be necessary to inform users of modification to the equipment.

Instructions: Detailed instructions for modification. Additional items required. Identification of products requiring modification.

9 Disposal instructions

Disposal, demolition etc. and hazards.

A more intensive study of this standard may be considered useful. Copies can be obtained from the British Standards Institution (address given under Useful Information), or from Her Majesty's Stationery Offices.

Part 2 of BS 4884 'specifies requirements for the layout* and preparation of technical manuals . . .' The standard recommends precise information for the layout of the front cover of the book, its spine*, title page*, and preliminary pages*. An example layout of a title page (as shown in the standard) is given in *Figure 1.2.*

11

Other topics covered include types and weight of heading, illustrations, references, production of camera copy* and indexes*.

Handbooks written for the domestic market usually fall into two categories: *User Guides* and *Maintenance Manuals*. They are often formatted for visual impact and present different problems to that of the military or industrially orientated document.

User maintenance is not encouraged by some manufacturers, even to the extent of voiding the warranty if certain

3.2(9) UDC 000.000

3.2(7) Submarine Aviation Board
Shrive Street, Shrift, SS1 2SS
Telephone : 000-00000 Telex 000000

3.2(4) Telecommunications
Technical Handbook
No. 1234 3.2(1)

3.2(8) General Editor
Manuel Footing

3.2(3) Radar Type XYZ 3.2(2)

3.2(5) July 0000

3.2(6) Anonymous Electronics Ltd
Wreak Street, Wroke, WW1 2WW
Telephone : Wroke 0000

Figure 1.2. Example of a BS title page (by courtesy of the British Standards Institution)

screws are tampered with. It must be admitted that there is often some point to this restriction, especially for modern solid-state electronic equipment, though one suspects that the withholding of information is mostly done for other reasons.

However, in writing any maintenance manual three distinct levels of maintenance and repair can be identified:

1 *Simple Maintenance.* This may include cleaning, replacement of discrete parts, oiling, simple test procedures, correction of minor faults.

2 *Technical maintenance on location with limited resources.* Of the type that a visiting television repair man might perform.

3 *Workshop technical maintenance.* Carried out by trained technicians with all necessary resources.

This threefold division illustrates a prime concept in the field of technical authorship: *level of readership*. All books must be written with the prospective user in mind. His level of skill and technical knowledge – not the author's – dictates the depth of treatment and even style of a technical handbook. In writing a maintenance manual this three-point division of skill and resources should be considered most carefully.

The User Manual or Guide is normally a straightforward operating instruction book containing some guidance on troubleshooting and a prominent referral to the repair network of the manufacturer or importing agent. It can range from a Hong Kong interpretation of a Japanese original printed on rice paper, to a sturdy paperback book written in a user-friendly style. Perhaps one should not judge the product by the quality of supporting documentation, but there is sometimes just a suspicion that a correlation exists.

Reports

A report is a document produced to convey factual information to a specific audience at a certain point in time. A report writer is usually someone with a special knowledge of the subject-matter, who may not necessarily be a full-time writer. He is quite likely to be an engineer, qualified either by his job status or training to undertake the particular task. Or he may be peripheral to the activity, but called upon nonetheless in an attempt to achieve objectivity.

Technical writers, especially those employed in the Tech. Pubs. department of a firm or organisation, are often asked to contribute to the company's report writing work load. There are several distinct advantages in this from the company's point of view. Such a writer can be expected to produce a result notable for its style and clarity. He will already have a good grasp of the technical aspects of the firm's production and so will not need an excessive amount of time at the briefing stage. And being slightly outside the drama of actual design and production, he should be able to resist internecine politics and influences.

A report, then, requires an element of factual objectivity, an appropriate style of writing and presentation, and a form which organises the material in the most accessible manner for the prospective reader. Its purposes are many, but the following six reasons for writing a report would cover most occasions:

- To give an accurate account of the subject.
- To present a basis for discussion.
- To arrive at a conclusion and plan further action.
- To make recommendations.
- To disseminate information.
- To provide a reference record.

A report can be subject-orientated – that is, divided into topics, or groups.

A report may be chronological – it represents a record in time, e.g. a progress report.

A report can be conceptual – dealing in abstractions or philosophical aspects of the subject.

From the technical author's viewpoint, we may say that reports are generated as part of a producer's internal information – his *Product Documentation*. This, in general, may consist of:

14

- An initial requirement, or technical proposal.
- A feasibility study.
- Design and research reports (project definition).
- Pre-production specifications: standards, performance, design philosophy.
- Evaluation documents: trials on reliability, maintenance, and so on.
- Ad hoc reports: investigation, progress, etc.

The requirement report, or technical proposal, begins the whole project. It sets out the requirement in general terms and allows the Board and their top technical advisers to make a viability decision on the proposal.

Once the go-ahead is given, a feasibility study will be commissioned in which all aspects, financial, technical and temporal are examined in detail. Solutions are arrived at, and time-scales suggested based on the best estimates then available. Subsequent stages, leading towards the making of a first model or prototype, involve even greater detail in producing a definition of the project in terms of design parameters and research and development plans.

A considerable amount of data will have been accumulated at this point, and the formalisation of design and test specification can be considered. Evaluation documents leading to reliability trials and consideration of maintenance needs will also be started together with many other reports on progress, processes and investigations.

Each organisation has its own way of doing things, and an author in Tech. Pubs., or an engineer at any level, will be expected to become familiar with the company's policy and methods. We shall, therefore, concern ourselves with a standard type of report, and extract only general information common to all such documents.

Our standard report will contain the following features:

1 Title
2 Contents
3 Aim of the document
4 Summary

5　Body of the report
6　Conclusions
7　A bibliography* and/or an index may be required in a large-scale report

These categories should contain:

1　*Title*: Not a fancy metaphor or double entendre, the title of a report should be sober, stating clearly, and in as few words as possible, what the document is about. If the report has an internal reference number, this should be printed here.

2　*Contents list*: Seldom necessary in a short publication, but is usually included as a matter of course in larger documents.

3　*Aim*: Why the report was written, its terms of reference and general purpose.

4　*Summary*: Many reports are so long and detailed that some readers do not have the time, inclination or know-how to tackle the whole document. The salient facts and a concise summary of the conclusions, if any, are therefore included before the main discussion.

5　*Body of the report*: Here we have the main discussion of the subject-matter. It may be organised according to a house style* or the preferences of the writer. British Standard BS 4811 *Report Writing* is a valuable reference for authors likely to be much involved in this area.

6　*Conclusions*: The author of a report will often be asked to make recommendations on the basis of his investigations. This is frequently the main object of a report. The conclusions should flow naturally from the main discussion and not introduce any new material.

Report writing is not the easiest of authorship tasks. It depends on a certain blend of talents: observation first and foremost, combined with organisational skills and a crisp and clear writing style which leaves no room for ambiguity or misunderstanding.

Technical articles

An article is a compact piece of writing with a central point or theme. The Concise Oxford Dictionary defines it as 'a literary composition forming part of a magazine etc., but independent.'

Articles tend to be shorter than they were in the more spacious days before television, the emphasis now being on 'fact' as opposed to the idea of a 'literary composition'. An author, therefore, must convey his message succinctly and in an interesting way.

The writing of technical articles is a relatively low-paid business requiring accuracy and organisation on the part of the author. Some degree of motivation and persistence is needed for success. Product and market must be tightly matched to have any chance of publication.

Technical articles range from pure journalism of the 'interesting angle' or 'scoop' variety, to highbrow scientific reports on new theories or experimental data. Bisecting these extremes there are articles written for the publications of professional bodies, companies, trades unions, and, with a little added flair, the trade papers and magazines. Periodicals like *Nature* and *Scientific American* concentrate on state of the art research, and are usually written by the scientists involved. *New Scientist*, on the other hand, presents informed articles by technical journalists in a more user-friendly manner. Magazines such as *Personal Computer* are user-orientated and welcome the views of software operatives in the micro field. Free trade papers, *Computing* or *Datalink*, for example, are written by staff journalists with a high degree of 'computer-literacy', and are also open to approaches from computer professionals. There are countless others. The *Writers' and Artists' Yearbook* or *Willing's Press Guide* may be consulted for a comprehensive listing.

Two points emerge in the breakdown of article writing:

● *Organisation*: The material should be constructed and presented in a way which allows the reader maximum access to the author's meaning and his data.

17

● *Readability*: The article should be written in such a way as to maintain the reader's interest to the end. Well-worn clichés and dialects of Latin, such as NASA-English, are to be avoided.

Most technical writing is written to specification. In practice this means that fairly rigid rules are laid down at the outset, and a house-style prevents any rush of blood to the writer's head. An article, by nature, is a personal product; it reflects the interests and personality of the writer. All good writers acquire a particular style. All good articles develop their own momentum.

Article writing, then, is best learned by making a thorough study of a wide range of published pieces, especially in the area for which the writing is intended. Let us consider briefly the elements of good article writing.

Organisation Good organisation means that the information content of the piece is unfolded logically, perhaps in a narrative or chronological form, to aid the reader's understanding and assimilation. For instance, it is usually right to move from overview to detail, from the general to the particular. If the data is sequenced in time, it helps to present it in chronological order. If the material is largely philosphical or conceptual in nature, the author must use some skill in the logical transmission of argument, as well as having a deep knowledge of the subject.

Readability In any discussion of readability we are confronted with many imponderables. Readability touches on style which impinges on taste. And taste is the least definable of human qualities. However, as a general principle, it is undoubtedly true that simple, living English is more readable than Latinised abstractions, no matter how impressive they may seem in certain quarters. Most good writers of English prefer the concrete Anglo-Saxon word to the Romanised equivalent, despite the preponderance of Greek/Latin derivations in scientific terminology.

Article writing has much in common with other forms of technical literature. Much of what follows in this book on the design, development and production of technical publications applies equally well to the composition of technical articles.

Technical sales literature

Sales literature is qualitatively different from other types of technical publication. For one thing it presents a distinctively glossy face to the world, and this face is a vital element in its message. Smooth, urbane, sophisticated, it is instantly recognisable for what it is – an attempt to maximise the sales of a particular product or service.

Writing technical sales literature (TSL) can be a fascinating part of an author's work, particularly if he has some creative imagination and an ability to generalise grandly using a minimum of fact. In many ways it is a sort of game, with the writer keeping one step in advance of the diminishing credulity of the reader.

For this reason, the best technical sales literature is often the most factual and straightforwardly presented. Simplicity, as in theatre, usually conveys the starkest and most believable effects.

In writing TSL an author normally works in close collaboration with a graphic designer. If the book is to be a glossy* brochure either the author or the artist may suggest a theme or motif. *Figure 1.3* illustrates a jigsaw motif which appears throughout a brochure selling technical authorship services.

As a first step, a set of roughs or visuals* will be produced by the artist to impress the client and give him an idea of how the finished work will look. Within these confines the author will determine his writing policy, his sales pitch, and the length of text available to him.

Naturally, the style will be bright and punchy, without becoming too conversational or convivial. The text should match the graphics* in overall approach and complement

19

them with information. It is something of an art. Not one that suits the temperament and talent of all technical writers. But one that requires experience, flair, a good technical background, and an expressive use of the language. It is the icing on the cake. Well paid, and well worth doing if it pleases you.

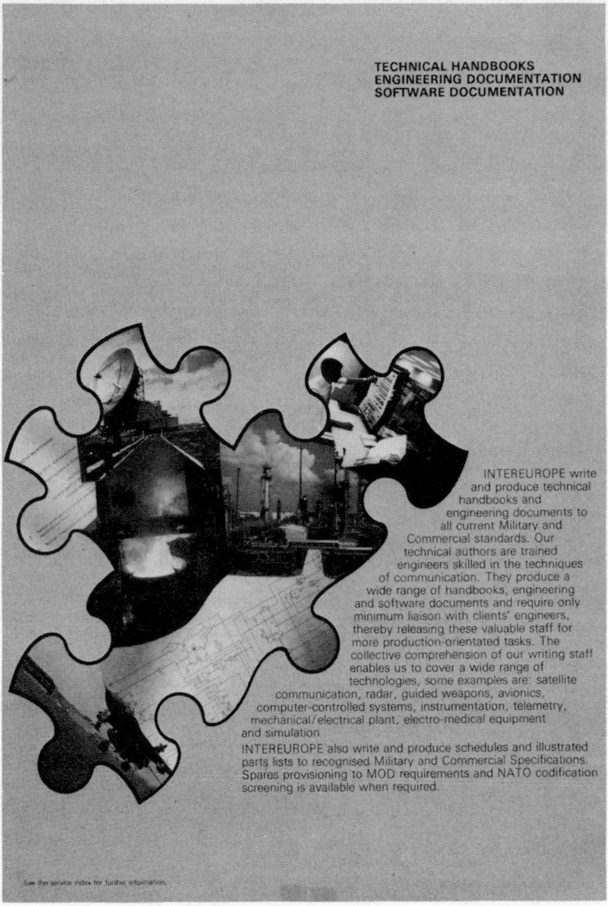

Figure 1.3. Jigsaw motif in a sales brochure (by courtesy of Intereurope Technical Services Ltd)

Technical training material

Fashion dictates most things, and educational training material is no exception. Many training courses nowadays appear in a form known as the Programmed Learning Package. This may be a mixed-media collection, containing audio-visual material, a film or video tape, cassette recordings and/or an accompanying book, or books. At the other end of the scale, it may come as a simple series of books or leaflets structured for layer-type or reinforcement learning.

Essentially, the aim of these packages is to administer a measured quantity of information in fixed units of time. Reinforcement is achieved by certain repeating techniques, gradually refining the data into a number of key concepts. The whole package is presented in a way that prevents even the most pampered student from becoming bored, and gives him a sense of having made progress at the end of each lesson.

A technical writer may be called upon to design and write Programmed Learning Packages, particularly training material for the engineering industry. Here we shall concentrate on the general concept and construction of such packages.

As for most tasks, the writer will follow a number of well-defined steps in designing his course. These will usually include:

1 **Learning objectives of the course**
   For example, on completion the student should be able to
   ● write a technical document to specification
   ● prepare camera copy
   ● proof read, as required
   ● amend issues, as needed

2 **Aids required**
   These might include, textbook, book of worked exercises, cassette tapes, and so on.

3 **Course coverage**
   The course will teach, for example,
   ● concepts of technical authorship

- how to design a technical document
- the writing and production of handbooks

4 Purpose

The purpose of the course may be to prepare the student for a career as an author, or for an examination such as the City and Guilds Technical Authorship paper.

5 Students and pre-knowledge

Continuing with our example, the course may be designed, as is this present volume, for students with a good technical background and knowledge of English who wish to write as a career or write the occasional technical document.

6 Course method/methods of study

The course could be linked to a series of lectures or demonstrations, or it may be intended for self study.

7 Course organisation

The author will probably find that his material falls naturally into a certain number of lessons. In the world of the Programmed Learning Package, these will be referred to as 'papers', or 'assignments' or, worse still, 'units'. Each lesson may contain

- textbook reading material
- reading from other reference sources/listening to cassette recordings/audio-visual presentation
- a series of exercises related to the reading phase and aimed at reinforcing the assimilated information
- additional exercises presenting unfamiliar aspects of the principles learned – perhaps a number of practical problems

8 Outline of each lesson

At this stage, enough material should be available to allow a breakdown of the content of each lesson, and some indication of the time required to study it.

Examples:

Lesson 1: Introduction to the course objectives: methods used and supporting material. Study time – 20 minutes.

Lesson 2: Introduction to the products of technical authorship. What an author does and how he goes about his work. Study time – 1 hour.

9 Define learning objectives of each lesson

The learning objectives are usually placed at the head of each lesson to give the student an aim during his study period.

10 Course length

A calculated estimate of the course length should be made, and guidelines laid down for prospective users.

This step by step systematic approach to the designing of a learning package can be similarly extended to the individual lessons. The same sort of breakdown can be used. Take our hypothetical Lesson 2, for example:

Lesson 2

Introduction to the products of a technical author. What an author does and how he goes about his work.

1 Overall view A general section encompassing the whole field to be studied in the lesson.

2 Learning objectives An assessment of what the student should achieve at the end of the lesson.

3 Materials required Props or other items needed.

4 Time required The time estimate for an average student to complete the lesson.

5 Body of the course The main body of information.

6 Summary of key concepts Basic elements of the material.

7 Review exercises Exercises reviewing the foregoing information.

8 New problems Exercises presenting different facets of the material such as might be encountered in real life.

Link to next lesson A lead-in to the next lesson to maintain continuity and highlight the progression of the data.

These are the essential elements required for the construction of programmed learning packages, and indeed, for most other types of course. Two aspects may now be considered in more detail: the audio-visual presentation and the educational textbook.

Presentations

Audio-visual work for industry can be placed under two headings:

● Information (including training)
● Sales (including publicity, exhibitions etc.)

AV presentations are often used for general staff training or for imparting information about a new system or equipment

Figure 1.4. Carousel projector (by courtesy of Kodak Ltd)

policy. If the organisation is large enough, an entire 'information package' may be commissioned, including films, audio-visual presentations, and a variety of supporting literature. In this section we shall look at the standard form of audio-visual presentation – a series of transparencies (slides) in a revolving Carousel projector (*see Figure 1.4*).

The slides are synchronised with the sound track on a tape recorder; slides are changed by a pulse recorded over the voice commentary on the tape. The writer's job, in most cases, is to write the supporting script using a predetermined slide series. Occasionally, he may be asked to produce the package from concept to presentation, a job involving liaison with a photographic studio. In the first case, a simple brief on the nature of the subject may be enough to complete the writing of the script. In the latter, some specialist knowledge is usually necessary.

For a small AV presentation, a colleague with a pleasant voice may be chosen to record the script. In a large company, the commentary will often be spoken by actors or professional speakers working in studio conditions.

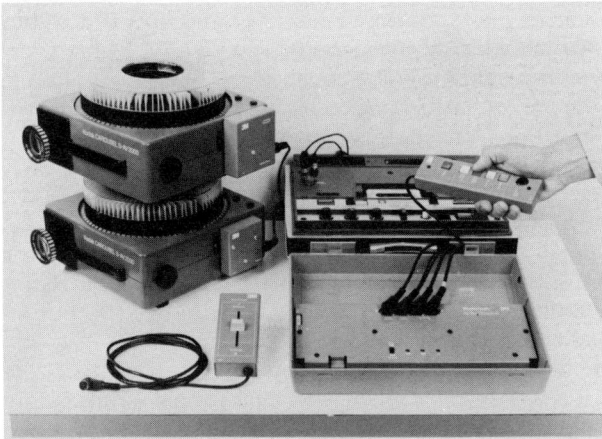

Figure 1.5. Audio-visual equipment (by courtesy of 3M Ltd)

But whatever the size of the job and the cost of production, the basic principles remain the same. The script is recorded on to cassette tape (reel to reel recorders are now rarely used). A backing, or music track, is added if desired, and a series of pulses are overlaid (*see Figure 1.5*). The sync. pulses, although not audible to the listener, automatically advance the visuals as the tape plays. They are recorded on to a separate track, so that the sync. can be erased or adjusted without affecting the speech track.

An audio-visual production, combining script with slides, should be approached in much the same way as any other writing task. The problem set, as always, is one of communication, and certain common features flow from this:

- The material should be fully understandable, and written to the level of expertise of the audience.
- A general overview should be followed by a progressive breakdown into relevant details.
- For conceptual material, abstract ideas may be illustrated by anecdote or concrete example.

A special point relating to audio-visual scripts is that they are by nature highly concentrated and equally, highly selective in terms of detailed material. There is an upper limit to the weight of factual matter that can be accommodated within a 15 or 20 minute commentary. A script that sounds like a list or a page from a telephone directory is hardly likely to be a good vehicle for its subject.

The answer is to paint with a broad brush, leaving much of the detail to the supporting literature, and maintaining a lightness of touch without losing sight of the main purpose – to complement the visual images in an interesting but informative way.

The layout of an AV may vary with house style (*see Figure 1.6*). It is worthwhile looking at several sample scripts before putting words on paper. For example, a script may be typed landscape* (longer edge horizontal), fastened by a staple at the centre of its upper edge. As each page is turned, a line drawing or print of the appropriate slide is shown on the top sheet, and the commentary on the lower.

Alternatively, the document may be arranged portrait* fashion (the longest edge vertical) with only a slide number and/or brief description to identify it. Usually the A4 page is divided into two halves vertically, with the slide ID on the left and the voice script on the right. This is most suitable for short, snappy descriptions. If, however, the explanatory material is lengthy, either the landscape approach or a two-page spread should be used to give it more space.

As for length, a reasonable rule of thumb is that 2000 words can be spoken in 15 minutes. Of course, long pauses will

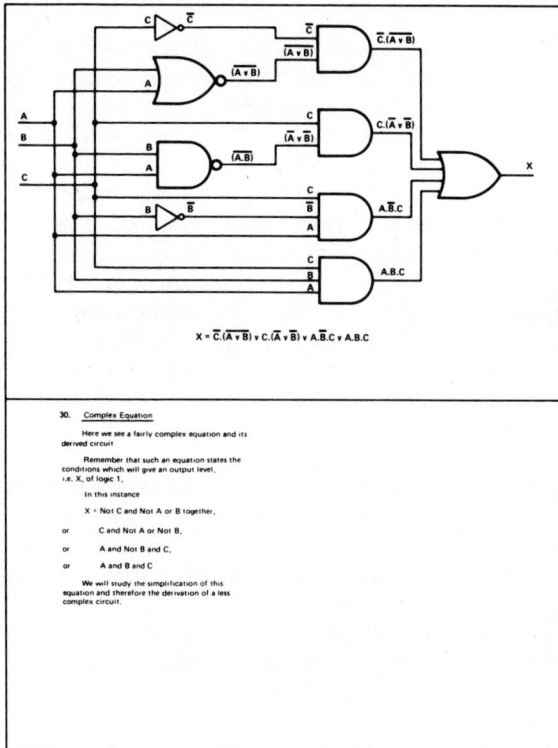

Figure 1.6. Example layout of AV script

reduce this, while a continuous conversational style will extend it by as much as 300–500 words. At all events it is advisable to read the script through at the projected pace with one eye on a stopwatch.

Educational textbooks

Books written specifically for educational purposes fall generally into two categories:

- Traditional textbooks for schools and colleges.
- Non-fiction books aimed at the general market.

This division is perhaps not so clear-cut as it used to be. Many textbooks are now packaged in a way that gives them some appeal in the bookshops. And conversely, the average non-fiction volume is seen as having an educational value in an age when information is at a premium.

Most technical books have some educational intent. An educational book should contain a body of knowledge in a format suitable for its comfortable assimilation. In practice many textbooks are opaque and some are unapproachable. This is probably because those with the greatest expertise may not necessarily be the best writers.

There is much to be said for the modern tendency, adopted because of the rising costs of book production, of applying some commercial values to all educational works. There is no reason at all why a textbook should not be readable, even by the general reader, and, depending on its subject matter, why it should not earn its keep in the bookshops.

For the writer who wishes to enter this field, the following definitive books should be read:

The Truth About Publishing, by Stanley Unwin.

Non-Fiction – a Guide to Writing and Publishing, by David St John Thomas.

Both authors have made their names as publishers as well as writers, and both give an excellent account of their subject.

Software documentation

It has been estimated that 70 to 80 per cent of the cost of developing a new computer system lies with the software (the programs and documentation that run a computer – anything not definable as 'hardware'). For over a decade, hardware costs have been falling dramatically as one leap in technology after another has come on to the market. Software overheads, in terms of man-hours per program, have hardly changed at all. When one considers that the mean time for developing a new computer application is about two years and the number of 'bugs' (errors) are large, affecting up to 5 per cent of program lines, it is scarcely surprising that program maintenance is an expensive business.

Computer programmers are notorious for failing properly to document the programs they write. And this is never realised until there is a need for program maintenance, or the programmer in question leaves the company bequeathing his eccentric codification to his colleagues. This is an unfortunate tendency because fully documented software contributes to reducing the continuing cost of program maintenance.

The software author or documentalist may become involved at any stage in the process of development. He may sit in with the programmer at the very beginning – an enlightened procedure now being adopted by a few companies – or he may be called in when the program has been written. In the worst case, he may be invited to participate when the actual programmer has departed, leaving behind a trail of poorly documented software.

He may be asked to document a single program, or a software system. The difference here is that a *program* performs a unitary function, whereas a *system* is a *set* of communicating programs.

The development of a program, or suite of programs follows much the same lines as any other technical product, including technical documentation:

- Requirement
- Specification
- System design
- Software system design
- Program design
- Implementation

- Integration
- Acceptance testing
- Maintenance, etc.

Figure 1.7 shows the combination of software and hardware documentation for any system.

Software documentation embraces the whole field of programming languages, while also taking in the disciplines of technical authorship. It exists in a curious interface between abstract ideas and concrete implementation. It deals with the 'intelligence' that makes a machine perform.

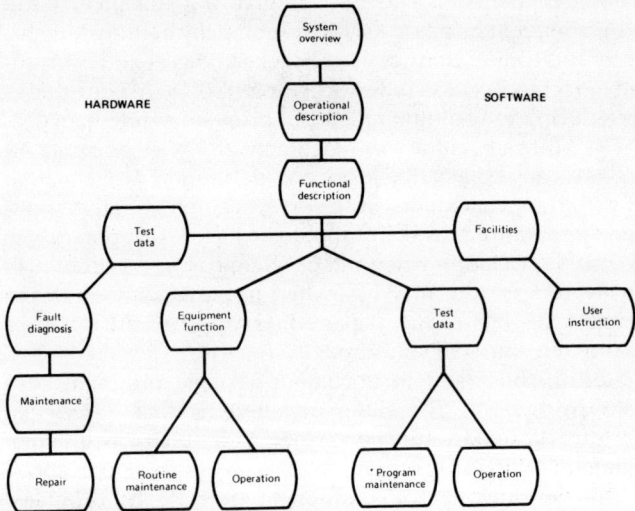

Figure 1.7. Soft/hardware documentation

Good software documentation will probably contain some or all of the following features. 'Probably' because as a fairly recent subject, no two practitioners will agree on the perfect requirement. The breakdown is therefore only a suggestion. But it will serve to give an impression of what software documentation is all about.

Title page (including reference number)
Contents
Outline of program
Update or amendment record
Program specification:
  Mathematical basis
  Program description
  Limitations and restrictions (presently known)
  Description of input
  Description of output
Programming information:
  Block diagram
  Specimen test data
  Program listing
  Specimen of output from test data
  Record of program development
  Operating instructions

Many of those whose job is writing programs will consider the documentation an irksome chore. There is a move therefore towards a concept called the 'self-documenting' program. The aim is to use to the full the annotating features of the programming language itself. The programmer then relies almost entirely on the program listing for maintenance and other purposes (*see Figure 1.8*).

Unfortunately, like most other attempts at improving software productivity, the concept is not as effective as it appears. There is at present no alternative to full program documentation. The following is a more detailed look at the above breakdown:

Title page
More than just a title page, it should bear a descriptive title of the program, references (if any), programmer/analyst's name, or names, and dates of specification and completion.

Contents
List of contents.

Outline of program

Brief description of the function of the program, language used, special techniques incorporated, system configuration written for, and utilities or operating system required.

Update record

Reason for, and dates of any amendments together with known effects.

Program specification

Mathematical basis   Any mathematical techniques used.

Program description   General description of program including loops and procedures.

Limitations and restrictions   Restrictions on field size, upper limits on data, and general accuracy.

Description of input   Input parameters or data requirements.

Description of output   Details of output.

```
780 PRINT"TOTALS",""ST,""SP
781 PRINT"------------------------------------"
782 PRINT"TOTAL RECEIPTS SATURDAY",""ST+SP
783 PRINT
785 INPUT"DO YOU WISH TO CONTINUE WITH SUNDAY";R$
786 IFR$="YES"GOTO799
787 IFR$="NO"GOTO2000
799 PRINTCHR$(12);
800 PRINT"TYPE NUMBERS OF GOODS LEFT SUNDAY"
801 INPUT"Cockles";E
802 INPUT"S.Prawns";E1
803 INPUT"Mussles";E2
804 INPUT"Whelks";E3
805 INPUT"Kippers";E4
806 INPUT"K.Prawns";E5
807 INPUT"J.Eels";E6
808 INPUT"Crab";E7
809 PRINTCHR$(12);
850 PRINT"SOLD SUNDAY","TO PAY","PROFIT"
851 PRINT"Cockles "D-E,V1*(D-E),R1*(D-E)
852 PRINT"S.Prawns"D1-E1,V2*(D1-E1),R2*(D1-E1)
853 PRINT"Mussels "D2-E2,V3*(D2-E2),R3*(D2-E2)
854 PRINT"Whelks  "D3-E3,V4*(D3-E3),R4*(D3-E3)
856 PRINT"Kippers "D4-E4,V5*(D4-E4),R5*(D4-E4)
857 PRINT"K.Prawns"D5-E5,V6*(D5-E5),R6*(D5-E5)
858 PRINT"J.Eels  "D6-E6,V7*(D6-E6),R7*(D6-E6)
859 PRINT"Crab    "D7-E7,V8*(D7-E7),R8*(D7-E7)
860 PRINT"------------------------------------"
861 K1T=V1*(D-E):K2T=V2*(D1-E1):K3T=V3*(D2-E2):K4T=V4*(D3-E3)
862 K5T=V5*(D4-E4):K6T=V6*(D5-E5):K7T=V7*(D6-E6):K8T=V8*(D7-E7)
863 L1P=R1*(D-E):L2P=R2*(D1-E1):L3P=R3*(D2-E2):L4P=R4*(D3-E3)
864 L5P=R5*(D4-E4):L6P=R6*(D5-E5):L7P=R7*(D6-E6):L8P=R8*(D7-E7)
865 GT=K1T+K2T+K3T+K4T+K5T+K6T+K7T+K8T
866 GP=L1P+L2P+L3P+L4P+L5P+L6P+L7P+L8P
880 PRINT"TOTALS",""GT,""GP
881 PRINT"------------------------------------"
```

Figure 1.8.   Computer program listing (by courtesy of W. Pethers)

Programming information

Block diagram   A general system diagram. It is not proposed to include a detailed flowchart here – although some documentalists do – because it is almost impossible to amend in line with program maintenance changes.

Specimen test data   It is not feasible completely to test every loop and procedure in a program since the variations may be nearly infinite. Test data are used within certain limits, however, and this, plus expected results should be listed at this point.

Program listing   A print-out of the program (*see Figure 1.8*).

Specimen of output from test data   Useful for debugging.

Record of program development   A log of progress during testing.

Operating instructions

These may appear as a separate user manual, or be part of an overall document.

Software and software documentation are developing quickly. New program development tools may even eliminate completely the need for applications programmers in the years ahead. In the present state of the art there is a steady demand for the careful software author. He must be a programmer, a writer, and a solver of conundrums. The job is demanding. But the rewards are substantial.

Summary

Manuals

Level of readership: Technical manuals are produced for the skill level of the reader/user. There are at least three levels for maintenance handbooks.

Customer support documentation: All literature written with the aim of providing a customer with accurate, up to date information about a particular product.

Specification: Detailed description of a product. A document prepared by an authority as a basis for the production of a technical manual.

BS 4884: British Standard specification for technical manuals. Uses a nine category system of book organisation.

Synopsis: A document, varying in detail, providing a summary of the contents of a book or document. Typically it contains chapter headings, subheadings and subject breakdown. It may also include a page count and/or an estimate of the number of paragraphs. A list of illustrations is usually provided. Useful for costing and organisation of material.

Vetting: A technical check on draft material.

Editing: A check on language, style and specification errors.

User friendliness: A concept which gives priority to the needs of the user. Often ignored by elements in the computer and electronics industry, but an excellent sales aid in a competitive market.

Reports

Product documentation: Literature which provides information on a product and its progress.

Purposes of a report: To give an account of a subject. To arrive at a basis for discussion. To reach a conclusion. To plan further action. To make recommendations. To disseminate data. To provide a reference or record.

Subject-orientated report: Divided into topics or subject groups.

Chronological report: 'Real-time' record of events or progress.

Conceptual report: Abstract or philosophical aspects of a subject.

Main features of a report: Title; contents; aim of report; summary; body of document; conclusions; index.

Technical articles

Article: Compact piece with a central point or theme.

Angle: Journalistic term relating to the orientation of an article in terms of its subject matter.

An article requires: Organisation – construction allowing the reader maximum access to the subject; and, Readability – quality of a piece of writing that maintains a reader's interest throughout.

Technical sales literature

Glossy: A brochure produced on high quality paper with coloured illustrations and which attempts to maximise the sales of a product or service.

Motif: A theme or recurring logo linking the pages of a book.

Visuals: 'Roughs' or visuals are a dummy copy of a brochure sketched out by the graphics artist.

Style of TSL: Style tends to go with fashion. Currently the approach is 'punchy', and 'snappy', and staccato, with phrases converted into sentences, and an air of underlying neurosis.

Technical training material

Programmed learning package: A collection of aids, usually 'mixed-media', structured for layer-type or reinforcement learning. The aim is to administer a measured quantity of information in fixed units of time.

Reinforcement learning: A spin-off of Pavlov's classical conditioning experiments on dogs, but this time aimed at humans. Reinforcement is achieved by repeating techniques using data which may be subtly veiled to make it more palatable to a modern audience.

Audio-visual presentation: A series of transparencies (slides) in a revolving Carousel projector, with the slides synchronised to a sound track on a connected tape recorder by means of inaudible pulses. Alternatively, an AV may be any form of presentation in which pictures and commentary are used together.

Length: Can be any length, but generally, anything over 20 minutes can lead to audience resistance. About 2000 words can be spoken in 15 minutes.

Educational textbooks: Traditional learning material produced exclusively for schools and colleges, or non-fiction
books aimed at the general market through bookshops.

Software documentation

Software: Anything in a computer system not definable as
'hardware' – the programs, operating system, compilers,
utilities and documentation.
Program maintenance: The act of 'patching' or 'fixing' a
program when a 'bug' is found. In practice, there is no way
of testing every loop and procedure in a long complex
program. Consequently, errors may be constantly dis-
covered in even the most mature software. The
documentalist annotates programs in such a way as to
provide maximum assistance for subsequent maintainers.
Stages of software development: As any other technical
product – requirement; specification; system design; software system design; program design; implementation;
integration; acceptance testing; maintenance, and so on.
Self-documenting programs: A program which is fully described in the listing using the annotation features embed-
ded in the language itself.

2

Design phase

Chapter 1 outlined the areas of documentation covered by the term 'technical writing'. It should be clear by now that the products of technical writing are so wide-ranging as to almost defy precise definition. In fact, if a piece of work involves factual material of a technical nature and is expressed largely in words, it is potentially a product of a technical writer.

This Chapter is concerned with the first phase in the preparation of any technical document, the *Design Phase*. The principles described here are common to all authorship products and to the design of almost any publication.

The flowchart (*Figure 2.1*) represents a series of steps on the way to an objective – an agreed synopsis – interrupted by a question which sets up a loop series in the event of a negative response. Therefore, if the synopsis* does not meet the requirement* or specification*, the author is directed around the loop and back onto the flow path for as many times as is necessary to arrive at his end-of-phase goal: final synopsis agreement. Each of the following phases – Development and Production – has a similar flowchart which is used as an aid in the evaluation and costing exercise.

Requirement

The first thing an author hears about any project is the requirement. It may be someone on the telephone saying:

'We have a requirement for an author to write a user manual for our new personal computer. We're looking for a

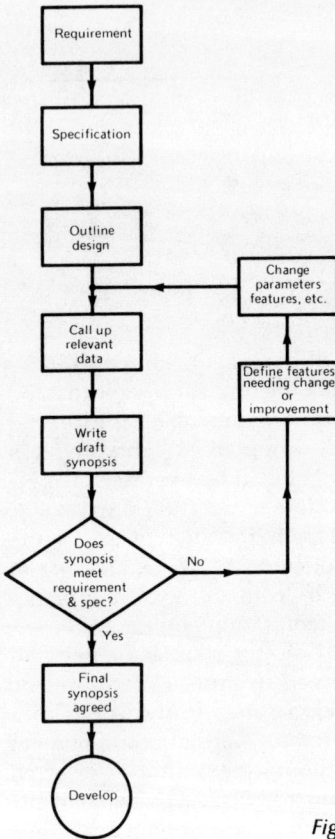

Figure 2.1. Design phase flowchart

glossy presentation . . . a very user-friendly text . . . lots of good illustrations . . . in fact, up-market and high class. We want typed draft only with author's sketches . . . We'll do the artwork and the camera copy . . . Oh, and we need the final draft in four months. Do you think you can handle it?'

That is a very explicit requirement. They are not always like that. Often a requirement is vague and undefined, either because the client does not know what he wants (but he will know it when he sees it!), or else he is deliberately pitching the ball into the writer's court on the basis that he (the writer) is the expert in these matters.

The first move, therefore, is to establish the requirement for the job. This is best undertaken by setting out a list of precise questions for the prospective client, omitting as little as possible. Writer's tend to have their own way of doing this, of course, but the following eleven questions will be found to be as good as any:

1   Is there a specification to be written to, or a house style covering the company's documentation?

2   What is the target readership? i.e., what depth of treatment will be needed?

3   Is there a deadline for the production of the manual? Can the author consult a project network?

4   What technical information is available? – reports, feasibility studies?

5   What is the maintenance philosophy? Are any diagnostic aids built-in to the equipment?

6   Are there subcontractors involved in the project? Is their documentation available?

7   Are commercial parts to be incorporated in the equipment?

8   Who are the writer's contacts – names, status, and telephone numbers?

9   How is the draft to be validated (vetted)? How long will this take?

10   Will there be editorial control? And who will have it?

11   Details of further documentation meetings.

Specification

When the writer has established the requirement in as much detail as possible without making a nuisance of himself, he should turn his thoughts to the specification. This area will cover the format, length and presentation of the document. Together with the requirement it will allow a detailed definition of the finished product. In most case a writer will be asked to write to a pre-existing spec, and to these we shall now turn for further developments.

A specification is a document prepared by an authority (industrial firm, ministry or standards institution) as a basis for the production of technical literature. We have already considered in some detail the specification for technical manuals produced by the British Standards Institution, BS 4884. The next step is to survey the rest of the field.

The list of specifications that follows contains those that are in general use, and some which are pending at the time of writing. Four of the specs listed have now been replaced, but are included here because currently used handbooks written to these formats may require updating or amendment. These are signified (0). For the first-time reader of this book a cursory glance at the list is all that will be necessary. Its main function, however, will be as a reference for future use.

British standards

BS 1219 (1958) Recommendations for preparation of mathematical copy and corrections of mathematical proofs: preparation of printing copy; printers' and authors' proof corrections.

BS 1991 Letter symbols, signs and abbreviations.

Part 1 (1976) General : Symbols for units and quantities widely used in science and technology.

Part 2 (1961) Chemical engineering, nuclear science and applied chemistry: symbols for quantities and abbreviations for units.

Part 3 (1961) Fluid mechanics: symbols for quantities and abbreviations for units.

Part 4 (1961) Structures, materials and soil mechanics: symbols for quantities and abbreviations for units.

Part 5 (1961) Applied thermodynamics: symbols for quantities and abbreviations for units.

Part 6 (1975) Electrical science and engineering: symbols for quantities and abbreviations for units.

Supplement No 1 (1973) to Part 6: List of subscripts for electrical technology.

BS 2517 (1954) Definitions for use in mechanical engineering.

BS 3527 (1962) Glossary of terms relating to automatic data processing: Standards and definitions for automatic data processing and computers.

BS 3939 Graphical symbols for electrical power, telecommunications and electronics diagrams (Section 1 to 30).

BS 4210 Specification for 35 mm microcopying of technical drawings.
   Part 1 (1977) Operating procedures.
   Part 2 (1977) Photographic requirements for silver film.
   Part 3 (1977) Unitised microfilm carriers.

BS 4884 Technical manuals.
   Part 1 (1973) Content.
   Part 2 (1974) Presentation.

BS 5261 Copy preparation and proof correction.
   Part 1 (1975) Recommendations for preparation of typescript copy for printing.
   Part 2 (1976) Specification for typographic requirements, marks for copy preparation and proof correction, proofing procedure.

BS5536 (1978) Specification for preparation of technical drawings and diagrams for microfilming.

The BSI Yearbook gives a complete listing of British Standards.

Civil air transport

ATA 100 Manufacturers' technical data (Airborne).

ATA 101 Manufacturers' technical data (Ground equipment).

ATA 200 Spares provisioning and procurement data.

Ministry of Defence (MOD) – Air

TPN (L) 114 General requirements for air publications.

AvP70 Air technical publications.

AvP77 Requirements for Anglo-French publications.

MOD – Army

AESP 0100-P-005-010 Specification for army equipment support publications.

(0) EMER General A022 Technical specification for the production of EMERs.

(0) EMER General A025 Physical standards for EMER production.

(0) EMER General A030 Preparation of illustrations for EMERs.

(N.B. EMERs have been replaced on new equipment by AESPs)

MOD – Navy

NWS1 Handbook for weapons systems and equipment.

NWS10 Development documentation system (See section on DDS under 'Miscellaneous Matters') for weapon systems and equipment.

DG Ships 607 Preparation and production of technical publications to Joint Services standards.

Joint Service Publications (JSP)

JSPs represent MOD's attempts at standardising documentation specifications across all three services. Eventually JSPs will be universally adopted (with certain variations) in Army, Navy and RAF publications.

JSP 181 Technical standards – General – Physical.

JSP 182 Technical standards – General – Presentation and layout.

JSP 183 Technical standards – Illustrations.

JSP 184 Technical standards – Classified publications.

JSP 186 Spec for illustrated parts catalogues.

JSP 187 Spec for illustrated parts catalogue for all equipment.

JSP 188 Documentation of software in operational real-time computer systems.

AQAP-11 NATO spec for technical publications.

American Department of Defense

MIL-M-7298C Manuals for commercial equipment.

MIL-M-38784 General requirements.

MIL-M-63000 Instructions for manuscripts and illustrations.

MIL-M-24100A Manuals, orders and other technical instructions for equipment and systems.

Central Electricity Generating Board (CEGB)

TP30 Power station instruction manuals.

CEGB 9861/1(7) Graphical symbols – Plant and valves.

CEGB 9861/1(9) Graphical symbols – Instrumentation for control purposes.

CEGB 9861/1(2) Graphical symbols – Electrical.

CEGB 098/16 SI units for operational purposes.

CEGB 98613 Plant numbering and nomenclature.

CEGB 9861/2 Letter symbols and abbreviations.

(0) DM 098/9 Commissioning – Operation – Maintenance procedures and instructions.

British Steel Corporation

CS30/SP/ST/TD0054 Instruction manuals: Steel-making equipment.

   This list is by no means exhaustive. It does, however, cover most of the specs in general use, and should prove of value as a general reference. As can be seen, specifications are usually designated by long strings of alpha-numeric characters followed by titles of Victorian prolixity. The beginner must take this in his stride. It is all part of the mystique of technical publications.

Outline design

Once the specification and requirement are known, a first attempt at outlining the whole project can be made. We call the result of this stage an 'outline design' because, while providing the basis for later events, it is not yet a complete definition of the task. It does, though, anticipate:

- The information the author will need.
- The synopsis of the document.
- The cost estimate of the work to be done.

The outline design summarises the content of both specification and requirement, and looks forward to the sources and type of information needed by the author in writing the book. As with previous stages we can put the outline design into the form of a number of questions:
   1   What is the purpose of the document? Customer information, sales confidence aid, training, reference or information for internal staff, product description or maintenance manual?

44

2   *What will be the format of the document?* Glossy publicity leaflet or descriptive brochure, summary card, reference/user/maintenance handbook, suite of manuals?

3   *How many copies are needed?*

4   *What is the intended readership?* Customers – actual or potential, company personnel, management, trainees, fully trained operatives?

5   *What are the sources of information?* Existing documents, training courses, practical experience, company personnel?

6   *How frequently will the publication be updated?* Never, occasionally, often?

7   *Who will do this, and how?* The author, company staff? By complete reissue, replacement of pages or sections, separate sheets for manual amendment?

8   *What form of presentation will be used?* Binding: 3 or 4 ring binder (plain or printed), lie-flat plastic (GBC), glued or stitched, multi-hole ring binder or welded plastic spine; contents: typed, word processed, typeset, full page or columns, pre-printed sheets (running head or company logo); illustrations: full or half page or in-text, half-tones* (photographs) – colour or monochrome, line drawings, flowcharts, block diagrams; cover: special artwork?

9   *What is the deadline?*

The answers to these questions, some of which arise from the specification and requirement, will make up the outline design of the product. Before writing a full synopsis, however, an author must survey the subject-matter in detail to determine its extent and content. Without a good feel for the topic he will find it very difficult to decide on chapter headings and further subdivisions of the book. The next stage in the process, then, is to consider in some depth the whole field of information gathering.

Sources of information

In these wild days of the so-called information revolution, data is everywhere and at a premium. Data means power, and data has gone democratic.

This section will consider information from three broad standpoints:

- ● Printed information available from libraries or other sources.
- ● Verbal information imparted by personal contacts.
- ● Visual information.

We shall also discuss information gathering in general, and conclude by examining the restrictions placed by law on sources of information, namely, copyright.*

Libraries

Many large companies maintain extensive libraries on technical subjects for the benefit of authors working in their Tech. Pubs. Departments. No private company library, however, is likely to measure up to all demands placed upon it, and it is often necessary for a writer to approach a large public library in search of information. This would apply even more to the freelance author or student, whose personal resources are considerably fewer.

There are two things an author needs to know in this respect: how to handle the sometimes arcane organisations surrounding huge collections of books, and what assistance is available to make a search for data more efficient and profitable.

To begin at the top with the National Library of Great Britain: the *British Library*. This rather forbidding institution is subdivided into departments or divisions, each devoted to specific fields of learning. Some, like the Department of Oriental Manuscripts, do not concern us here, so we shall concentrate on the more technical areas. Full addresses of the following libraries are given under 'Useful Information'.

The *Science Reference Library*, formerly the National Reference Library of Science and Invention, is divided into two branches:

- ● The Patent Office Library, Holborn, London containing a unique collection of books and journals orientated towards industry.

● The Bayswater Branch exhibiting mainly books and journals on the life sciences.

The *Research and Development Department* of the British Library takes in an earlier and similar organisation. The Systems Development Branch is in Kingsway, London.

Another useful department is the *Lending Division* at Wetherby, Yorkshire, which encompasses the former National Lending Library for Science and Technology. This admirable institution offers a swift and comprehensive lending service accessed through other public libraries.

If, as is usually the case, the research is not weighty enough to justify lengthy sabbaticals browsing among the splendours of the British Library, it is often possible to find, or gain access to, the appropriate source of information at your local library, especially if you live in a city or large town.

Several specialised information services, available mainly by subscription, are also worth bearing in mind:

CICRIS at Acton Public Library is an amalgam of college, industrial and municipal libraries in West London.

HERTIS at Hatfield College of Technology is available to industrial and research organisations by subscription.

HULTIS at Hull Public Library.

NANTIS at Nottingham Public Library.

LETIS at Leicester Public Library.

LADSIRIAC at Liverpool Public Library.

ASLIB is the Association of Special Libraries and Information Bureaux, and is a national organisation supporting local branches. The ASLIB directory is worth consulting as a guide to the location of special reference material.

Before tackling the various classification systems adopted by libraries the world over, it is worth considering a logical approach to the search for an area of information, or specific technical publication, using the public library system.

The preliminary step in any search is to establish the general principles of the subject to be researched, and to identify the areas of particular interest. It is helpful at the start to become familiar with the technical jargon used by insiders to the subject. Despite protests to the contrary, all subjects have their secret codes, distinguishing initiates to the brotherhood from outsiders, such as prowling technical writers.

A first approach, therefore, must be to the technical dictionary (*Chambers's*, for example). This may be supplemented by excursions into the *Oxford Dictionary* or *Encyclopaedia Britannica*. These works can all be found on the shelves of the reference sections of most public libraries.

When the subject areas and terminology have been defined, the next stage is to scour the library catalogues noting all topic references and classifications. (The Library Association publish a useful volume called *Guide to Reference Material* by A. J. Walford and L. M. Payne.)

It should now be a straightforward matter to look through the classified index for details of relevant publications. A further move in this direction may be made by consulting:

British Books in Print.

Subject Guide to Books in Print (USA).

Should a library not stock a particular publication, it is often possible to ask the librarian to obtain a copy from elsewhere. This is a most useful service, and indeed in any complex search it pays to talk to the librarian who will have a great deal of experience in this area. Copying facilities are now available at most establishments, though the laws relating to copyright must be strictly adhered to. (Refer to section on Copyright.)

Two books which merit consultation for bibliography information are:

The British National Bibliography – which is a weekly list of books published in Britain with a monthly cumulative index.

The Cumulative Book Index (USA).

Periodicals are essential reading if you are dealing with recent data or state-of-the-art equipment. Periodical indexes include:

Ulrich's Periodicals Directory (R. R. Bowker, New York).
Periodicals – Guide to Reference Material.
British Technology Index.

Abstracts* (or classified extracts from periodicals) should
also be consulted. They include:
Electrical Engineering Abstracts
Chemical Abstracts
Metallurgical Abstracts

Trade publications and specialist journals may be found in:
Willing's Press Guide
Newspaper Press Directory
Guide to Current British Journals

For information on research organisations working on any
specific subject, consult:
Directory of British Associations
Trade Associations and Professional Bodies in the UK
Technical Services for Industry

Several books are published giving reviews of innovation
in specialist fields. The publishers of this present volume,
Butterworths, produce a handbook called *Progress in Auto-
mation.* Books serving the same purpose are:
Advances in Applied Mechanics (Academic Press)
Annual Report on Progress in Chemistry (The Chemical
Society)

Copies of any appropriate British Standards are often
invaluable sources of data. Two publications are useful in this
regard:
British Standards Yearbook and *Supplements.*
Library of Foreign Standards

The BSI also publish sectional lists of standards, which are
supplied free of charge upon application to the BSI Sales
Office (Address under 'Useful Information'). Examples of
these include:

Aerospace Materials and Components
Chemicals, Fats, Glues, Oils, Soap, etc.
Electrical Engineering
Machine Tools

Access to patent information may be obtained via the Patent Index or from the Patent Office.

Any large collection of books requires a method of classification. Without this, the task of finding any one publication, or a number of publications dealing with the same subject, would be almost impossible.

Three systems of classification are used:
- The Universal Decimal Classification.
- The Dewey Decimal System.
- The Library of Congress System.

The **Universal Decimal Classification** (UDC) has ten main divisions:

0 Generalities.
1 Philosophy, Metaphysics, Psychology, Logic, Ethics and Morals.
2 Religion, Theology.
3 Social Sciences, Economics, Law, Government, Education.
4 Philology, Linguistics, Languages.
5 Mathematics and Natural Sciences.
6 Applied Sciences, Medicine, Technology.
7 The Arts, Recreation, Sport, etc.
8 Literature, Belles Lettres.
9 Geography, Biography, History.

As far as the range of subjects relating to technical authorship is concerned, we can concentrate almost exclusively on division 6. If, for example, we are particularly interested in telecommunications, the sector subdivides thus:

6   Applied Science; Medicine; Technology.
62   Engineering Sciences.
621   Mechanical & Electrical Engineering.

621.3   Electrical Engineering.
621.39 Telecommunications, Radio, TV, etc.

The catalogue index will list all available titles on telecommunications under the designation 621.39.

The **Dewey Decimal Classification** contains a similar ten-point subdivision:

000 General.
100 Philosophy.
200 Religion.
300 Sociology.
400 Languages.
500 Pure Science.
600 Useful Arts (including Applied Science).
700 Fine Arts.
800 Literature.
900 History.

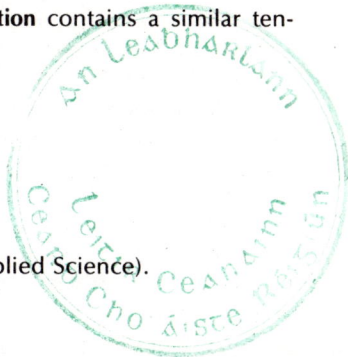

These ten classes are subdivided into ten divisions, and each of these into ten sections. The whole system thus totals 1000 sections, numbered from 000 to 999. By the addition of a decimal point, each section is further divided into ten. The process may be continued to any practicable length.

The **Library of Congress Classification** is an American system. It uses an alpha-numeric notational arrangement, which is not completely satisfactory since it lacks a commonality of class notation. The system was developed, as its name implies, to classify the publication stock of the US Library of Congress.

The **International Standard Book Numbering system**, commonly referred to as ISBN, originated in Britain in the late 1960s. It it run in the UK by the Standard Book Numbering Agency (address under 'Useful Information'). While not a legal requirement, it has been almost universally adopted by publishers, booksellers and librarians, and is extensively used in computer and tele-ordering systems.

ISBNs are allotted to books by publishers, who may have their own list of numbers, or by the agency itself. The number usually appears on the history page, and on the back cover of paperbound books. 7000 ISBNs are allocated every year.

The agency describes the number's role as identifying '. . . one title, or edition of a title, from one specific publisher . . .' It is unique to that title or edition.

With the introduction of Public Lending Rights, whereby authors will receive some return on books lent out at public libraries, it is just feasible that ISBNs will at some future date be made compulsory.

For a technical author libraries exist to provide information. Armed with an understanding of the main classification systems, and an idea of what to look for, no subject, no matter how arcane, is beyond reach.

We shall now look in more detail at three other aspects of 'calling up relevant data' – contacts, meetings, and information gathering in general.

Contacts

Once a writing project is under way, and the author has assimilated his technical brief through the customer's requirement and specification, his next move should be to establish lines of communication with the people most intimately involved with the day to day development of the project.

It may well be that an adminstrative person is put in charge of all contacts between author and engineers, and in this case all arrangements should be handled through his office. Alternatively, the chief design engineer or a project leader in charge of development work is the contact. If, however, the author has been subcontracted by the Tech. Pubs. Department of the company concerned – for whatever reason – it is usual to go through the head of this office, or his nominee.

For authors working permanently within this department, such contacts should present less of a problem.

At the outset of a writing task it is usual to make up a list or card index of all useful names. This file saves a great deal of time and trouble later, especially if it also includes information on job status, position in the hierarchy, and importance to the project development. Obliging contacts can be vital at awkward moments when, as happens in any project, things seem to be slipping away in a miasma of confusion.

It is of course a matter of courtesy – and common sense – not to approach technical staff until the brief has been fairly well mastered and the author can speak to a project engineer on something approaching equality. The best time to make effective use of contacts is usually high up on the 'learning curve'.

When arranging such meetings or site visits, due regard should always be paid to company procedures and security checks – now endemic in all organisations. Appointments should therefore be made through the correct channels.

Every visit should be approached in a careful and professional manner. This will probably involve:

● Listing areas which need clarification.
● Compiling a list of queries on modifications or changes.
● Making a note of questions of technical interpretation.

A good liaison visit will accomplish the following:

● Points answered.
● Technical queries clarified.
● Equipment examined.
● The way paved for the next stage of work.

Of course, if an author wishes to examine any equipment, advance notice must always be given to the people concerned so that the right hardware can be put in the right place, and the right engineer made available at the right time. Simple courtesy allied to forward thinking can always be expected to carry the day in these matters.

Meetings

Technical authors are frequently asked to attend, or even chair, formal meetings. A brief résumé of the rules governing business meetings will be appropriate here.

The chairman is the ruling authority at any meeting. It falls to him to make the initial arrangements and to draw up an agenda. His main considerations will be:

- Is the meeting absolutely necessary?
- Who needs to come?
- Are they all available on the proposed date?
- What is the precise subject to be discussed?
- What will it achieve?
- At what times will it start and finish?
- Where will it be held?
- What information is required in advance?
- Are any other items needed, i.e. projector, tape player or video recorder?

The next step will be to draw up an agenda. He will consult the various people involved on the time and date, the convenience of the venue, and on any topics they may wish to raise. The agenda will usually contain:

- Place, time and date of meeting.
- Subject, or subjects, to be considered.
- Subject order for discussion.
- Other points of interest.

The agenda should be distributed in advance to all the proposed attendees at the appropriate time, i.e. neither too early, nor too late. The ideal time for distribution is not so far in advance of the gathering that the people may forget, and yet giving them sufficient time to assimilate any brief and do all the necessary homework. At the meeting the chairman will:

- Start on time unless there are pressing reasons against it.
- Introduce newcomers.
- State the purpose and aims of the meeting.
- Follow the agenda as written.
- Let the meeting flow if progress is being made.
- Sum up the arguments if they are being lost.
- Pass on to the next point if the meeting is getting bogged down.

● Not allow awkward characters to dominate the discussion.
● Conclude the meeting on time if possible.

Meetings are useful in that they get people together face to face. Prevarications can be quickly worn down. Misconceptions, or areas not well defined, can be discussed and conclusions agreed there and then. On the other hand, a badly handled or mistimed meeting may just be a waste of everybody's time and effort.

Information gathering

The problems associated with technical information largely revolve around:

● Accumulation.
● Retention.
● Accessibility.

The human brain alas is not ideally organised for memorising complex blocks of data. Consequently, technical information should be referenced, and stored appropriately, for ease of recall.

A client will present information to an author in three ways:

● *Printed*: requiring referencing and filing.
● *Verbally*: requiring notetaking or tape recording.
● *Visually*: requiring sketches or photographs.

Printed information: writing jobs often fall into two categories – those which deluge the writer with ream after ream of printed matter, and those which grudgingly part with a few sheets in an engineer's indecipherable script. Naturally, whatever is given is gratefully received and filed accordingly. Printed information may come in the form of:

Manuals written for previous or similar designs.
Sales documents.
Specifications.
Test schedules.
Parts catalogues.

Diagrams – system diagrams, schematic flow drawings, circuit and layout diagrams, interconnection charts.

An ability to read and interpret such technical diagrams is part of the background expected of a technical author.

Manuals written for previous or similar models should be looked at mainly as a style sample, taking in such pointers as drawings, depth of treatment and other policy areas. On the other hand, it may be possible to identify sections which are common to the current manual.

Sales literature must be treated with some caution. It may occasionally prove useful in checking performance standards, and high definition photographs are often valuable in providing a visual record of the equipment.

Specifications and test schedules are another vital source of information and much data of a tangential nature may be extracted from them. Similarly, parts catalogues or items lists are essential documents in the writing of a maintenance manual. They should present an accurate, indexed list of all the component parts applicable to the system. An illustrated parts catalogue is even more informative in that it depicts the equipment as a series of keyed exploded views with each component fully annotated.

Diagrams, of one sort or another, represent a highly concentrated source of technical information. It is assumed that the initial background of any author will have brought him into contact with many documents of this type and so we shall not dwell too long on them here. The most primitive and therefore troublesome of these is the engineer's sketch. This is usually hastily scrawled on the nearest laminated object, the inside of a matchbox or a till receipt from Sainsbury's. Circuit components appear in a form of shorthand unknown to British Standards, and individual lines often meet severally in odd places. This is the pictorial equivalent of the anagram and should appeal to crossword aficionados with a low awareness threshold of profit margins.

System and flow diagrams give an overall assessment of a technical situation, revealing the relationships between components or functions in a system.

Circuit diagrams contain a series of graphical symbols drawn to show the sequence of events in a circuit and the component interconnections. It is largely a theoretical arrangement depicting functional information; and some components, such as integrated circuits, may be split up by function, the parts appearing in different areas of the drawing.

Layout diagrams represent the actual constituent components by shape and position in the design. A further extension of the circuit and layout drawings is the interconnection diagram – sometimes called a 'pianola'. This is a tabulated presentation of all the component interconnections within the equipment.

All printed matter should be suitably arranged and filed. In practice, most authors have their own personal system of filing, and manage quite well with it. The only proviso to this is that colleagues should not find the method too hard to crack if the author is, for any reason, indisposed.

Verbal information: Verbal information will usually form a sizeable part of the data input to any job concerned with state of the art equipment. Little, if anything, may have been written down at the early stages of development, and what has been committed to print may be out of date or misleading. A writer should prepare himself for this common situation by adopting a suitable system of notetaking.

Notetaking is something of a learned skill. It demands a systematic approach combined with the ability to extract the salient points from a flow of verbal information. Journalists develop this into a fine art. Few, if any, technical writers use shorthand, but many have devised for themselves a system of speed writing for clipped, yet accurate, notetaking.

An alternative, widely used, is the small, personal cassette recorder. Many companies now provide these machines for their writers and executives.

However, if a writer intends using a recorder to tape a conversation with an engineer, for example, it is a wise move to ask permission first. Engineers occasionally go dumb in the presence of a microphone, and executives have been known to wax philosophical while the tape supply dwindles away. Some people even resent a tape recorder as an intrusion of privacy.

Failing an opportunity to obtain a recording from the horse's mouth, the next best option is to record the tape yourself immediately after the interview while it is still fresh in the memory.

Visual information: We will assume that all photographs to be used in illustrating a handbook will be taken professionally. This is not to say that many writers are not also good photographers, but generally, their time is more profitably spent in attending to writing tasks. There are times, however, when a writer will almost certainly be called upon to photograph equipment for information purposes, or to supply an illustrator with the basis for a line drawing. Consequently he will need to possess the raw minimum of photographic skills. Nowadays, most companies employing writers who go on-site to obtain information will provide an elementary sort of Polaroid camera for collecting visual information. It must be said though, that while they have their uses elsewhere, they are hardly ideal machines for close, high-definition work.

A useful addition to any writer's baggage train is a good quality 35 mm single lens reflex camera. It should be capable of focusing down to about 45 cm (18 in) and have a maximum aperture of at least f/1.8.

Photography in a technical age has largely been reduced to following the instructions in the camera manual. Most new cameras are fully, or partly, automatic, and very little choice is left to the operator. Apart from focusing – made remarkably easy in a through-the-lens camera – the photographer has only to make a choice of either shutter speed (in a shutter-priority model), or aperture (in an aperture-priority instrument).

A light reading for a shot is expressed as a trade-off between two associated parameters:

- ● f/stop – the ratio of lens focal length (on infinity) and the diameter of the lens aperture.

● Shutter speed – the time, expressed as a fraction of a second, for which the shutter remains open.

For a given light intensity, opening the aperture wider (say, from f/5.6 to f/4) will require a corresponding decrease in shutter speed (say, from 1/125 th sec to 1/250 th sec). The difference between f/4 and f/5.6 is said to be one f/stop, in that f/4 allows twice as much light to pass through the lens as f/5.6. Therefore, by halving the time for which the shutter is open, one restores a correct exposure.

As an example, the following combinations are all correct for one specific light reading:

f/1.4	1/1000 th
f/2	1/500
f/2.8	1/250
f/4	1/125
f/5.6	1/60
f/8	1/30
f/11	1/15
f/16	1/8

Given the choice of these eight combinations, the most appropriate selection depends on the nature of the two parameters. A shutter speed should always be selected in relation to the subject and its capacity for movement within the field of the shot. For example, a handheld photograph will probably exhibit camera shake at speeds lower than 1/60 th or 1/30 th of a second. For speeds lower than 1/30 th, a tripod or firm resting surface is essential. The movement of the subject is another factor. A galloping horse will require at least 1/500 th and ideally, 1/1000 th of a second. A stationary vehicle, on the other hand, may be shot handheld at speeds as low as 1/30th, and at any speed on a tripod.

The aperture of a lens at the time of shooting presents the photographer with another problem of choice. Apart from considerations of light, the wider the aperture the smaller is the depth of field. This means that with the lens focused on a particular plane, at a specific distance from the camera, the range of tolerable definition before and beyond that plane increases as the aperture is decreased. At full aperture (say

$f/1.8$), very little depth of field is available. This may be useful for eliminating unwanted or fussy backgrounds, but it not so welcome when photographing a subject which recedes considerably in distance and which must be sharp overall. At $f/22$, however, depth of field on a standard lens is such that if the lens is set at 10 or 15 feet, almost everything from about 6 feet to infinity will appear acceptably sharp.

Going back to our list of combinations, $f/5.6$ at 1/60 th will give good handheld definition with reasonable depth of field, and would probably be the choice in a majority of situations. The modern camera offers that choice in two ways: shutter-priority and aperture-priority.

In the former, the aperture is set by the auto-mechanism on the basis of the light reading and the choice of shutter speed. In an aperture-priority camera, the aperture is set and the machine selects the corresponding shutter speed. There are fully automatic models with 'programmed' controls, but for technical work it is felt that some manual choice should be left to the operator.

Copyright

Under UK law any work expressed in writing or print is afforded legal copyright from the moment it is written. There are no formalities involved – no complex registrations, fees payable, and no licence of any kind required. As such it is one of the rare gratuitous benefits of authorship.

Copyright is vested in the author of a written work from the time of its completion until 50 years after the end of the year in which the author dies. Protection is not granted to ideas, however original, or themes and plots, however unusual, but to the actual words as written. The proviso is that the work is of sufficient length to have occasioned some labour or skill in its productions.

Copyright is the exclusive prerogative allowed by law to an author, or his assignee, to print, publish, or sell copies of his original work as and when he wishes. British law sets the following time limits on this legal protection:

- From the time the work is written until 50 years after the calendar year in which the author dies.

- For works originating from more than one author, copyright extends to 50 years after the death of the first to expire, or during the lifetime of the author who dies last, whichever is the longer.

- Photographic material enjoys copyright protection for 50 years from the end of the year after first publication.

- Artistic work and cartography have protection for the same period as for literary compositions.

Once a period of 25 years has elapsed, a work may be published, on payment of appropriate royalties to the author, who must be given adequate notice of publication.

Copyright subsists in the actual substance of the writing rather than its subject matter. It is how the meaning is expressed which is granted legal protection, not the ideas being conveyed. Having said that, it is still possible to infringe an author's copyright by paraphrasing or even summarising his work without permission, especially if it is apparent that the spirit of 'fair dealing' inherent in the Copyright Act has been exceeded. Short extracts, or referrals, are allowed, but any substantial quotation requires the writer's agreement and an appropriate acknowledgement in the work.

For a more lengthy discussion of copyright law, the reader is referred to:

A User's Guide to Copyright by M. F. Flint, Butterworths, 1979.

Copyright by K. Ewart, Cambridge University Press, 1952.

Alternatively, a useful article on British and American copyright appears in the *Writers' & Artists' Yearbook*.

The synopsis

A synopsis is a vital part of the design of any document. If it is carefully thought out and constructed, it will define the topic breakdown chapter by chapter, the amount of detail to be covered, and the number and type of illustrations to be included. It may also be possible at this stage to make an accurate estimate of the size of each chapter and therefore of the size of the book.

The synopsis is an essential preliminary step in the costing exercise. In conjunction with a suitable work schedule, which will allocate the time elements of all concerned, a detailed synopsis will define the project in its entirety before a single word has been committed to paper.

There are a variety of ways of composing a synopsis. It is really a matter of taste. Some writers prefer a narrative approach, so that the details appear as part of a block of text. Others adopt a tabular system. Whichever method is used, the essential data to be recorded for each chapter or section includes:

- Chapter or section number.
- Provisional title of chapter or section.
- Subject of chapter or section.
- Topic breakdown – subheadings etc.
- Illustrations – number and type.
- Page estimate.
- Other relevant remarks.

Some companies encourage a paragraph by paragraph system, in which a complex system of paragraph numbering is used, with extensive referencing. In this case, the number of paragraphs should be estimated.

The simplest way of setting out a synopsis is shown in *Figure 2.2*. A tabular presentation has the advantage of consistency, forcing the author to be specific. It is probably better suited to longer works than a simple narrative approach.

If a specification is used, such as BS 4884, the category breakdown recommended will determine much of the shape of the synopsis. Within each category, a more detailed consideration of the appropriate material will still be necessary.

CHAPTER No.	CHAPTER TITLE	SUB-HEADING (DETAIL)	ILLUSTRATIONS	COMMENTS
1.	TECHNICAL WRITING	MANUALS	2 LINE	
		REPORTS	—	
		TECHNICAL ARTICLES	—	
		SALES LITERATURE	1 HALF-TONE	
		TRAINING MATERIAL		
		Presentations	2 HALF-TONES	
		Textbooks	1 LINE —	
		SOFTWARE DOCUMENT-ATION	2 LINE	TOTAL
				3 HALF-TONES
		SUMMARY OF CONCEPTS	—	5 LINE
				PAGE ESTIMATE: 33
2.	DESIGN PHASE	REQUIREMENT SPECIFICATION	1 LINE	
		OUTLINE DESIGN	—	
		SOURCES OF INFO	—	
		Libraries	—	
		Contacts	—	
		Meetings	—	
		Info gathering	—	
		Copyright	—	
		THE SYNOPSIS	1 LINE	TOTAL
		THE WORK SCHEDULE	2 LINE	5 LINE
		COSTING	1 LINE	PAGE ESTIMATE: 40

Figure 2.2. Tabular synopsis layout

Once a synopsis has been written it should be circulated to all interested parties for agreement before proceeding to the next level.

During the writing of a book, it may not be possible to adhere rigidly to the initial outline, particularly as the idea matures, or if the subject matter is liable to design changes.

But if the synopsis is well thought out, and a certain latitude built-in at the start, it will provide a firm basis on which to construct the entire project.

The work schedule

One of the best ways to begin preparing a work schedule is to study the three flowcharts given in this book (Design, Development, and Production) and estimate the time likely to be needed for each step in the process (*Figure 2.3*).

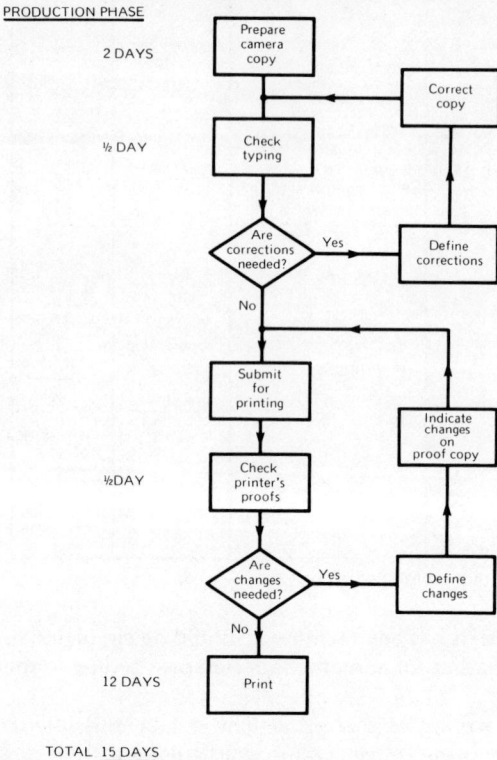

PRODUCTION PHASE

2 DAYS	Prepare camera copy
½ DAY	Check typing
	Are corrections needed?
	Submit for printing
½ DAY	Check printer's proofs
	Are changes needed?
12 DAYS	Print

Correct copy

Define corrections

Indicate changes on proof copy

Define changes

TOTAL <u>15 DAYS</u>

Figure 2.3. Time estimating on flowchart

The main variable will be the number of times an author has to go through each loop. Experience should give some indication here, but if not, it is certainly a wise practice to build in a contingency for perhaps two or even three trips around the course. An alternative would be to set a limit to this procedure, in agreement with the client, ensuring that some sort of draft is produced to time.

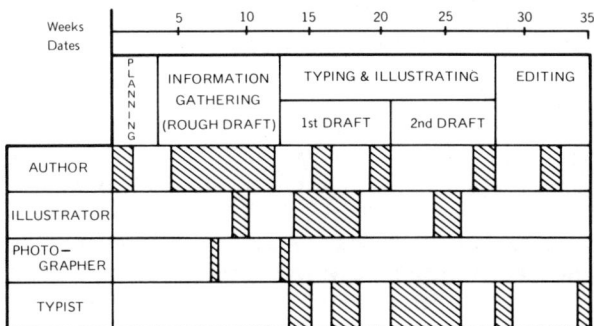

Figure 2.4. Time-based chart

A straightforward time-based chart may then be constructed, as depicted in *Figure 2.4.* For large-scale projects involving a number of volumes, the Project Evaluation and Review Technique (PERT) or the Critical Path Method (CPM) may be used. A discussion of 'network analysis' as these methods are called is given in 'Miscellaneous Matters'.

Costing

Costing is beset with problems for an inexperienced writer. Not only do costs of all kinds change unrelatedly over short periods of time, but many hidden variables surface only when the job has been finished and paid for.

Generally, a technical writer will not 'talk money' to a client. He will, however, be expected to 'talk time', and this, in many cases, is the major element in the final quotation.

The exceptions to this rule, who do indeed talk money to the customer, are senior writers with management responsibilities, and the freelance, who will need to do most things for himself. A writer publishing commercial books will of course negotiate a royalty on sales, or dispose of his copyright for a set fee.

Writing tasks are usually undertaken on the following terms:

Fixed price quotation
Cost plus
Limit of liability

Fixed price terms are most common for small to medium jobs where any overshoot of costs will not present too many problems. The advantage from the client's point of view is that he knows what he will have to pay out at the end of the job. For the writer, the main advantage is that, if the work can be completed inside the proposed timescale, the final profit will represent a larger percentage of the overall price package. Naturally, in exercises of this kind, it is vital to estimate author-time most accurately. Other overheads – typing, illustrating, material costs, and so on – are much simpler to arrive at, but where a writer runs into difficulties, for whatever reason, the usual 20 per cent contingency on his effort is soon eaten away, and the profit margin diminishes progressively.

The usual practice in fixed price situations is for the writer (or writers) to estimate the time required to complete the task, and also the number of liaison visits that may be needed. This information is passed to an administrator, or writer with experience of costing, who will add a variety of resources and materials, office overheads, time spent by supporting staff (typists, illustrators, and so on) and a standard profit percentage, to arrive at a preliminary figure. To this he will aggregate around 20 per cent for contingencies, and, in most cases, a percentage for VAT.

The figure will be appended to a document representing the quotation, in which all the terms of the job are set out in detail for agreement by both parties. This is regarded as a legal document in the event of a dispute, so it must be worded accurately and carefully.

The quotation is then sent to the client, who is, of course, at liberty to reject it and go elsewhere. If accepted, the job commences on an agreed date, and, at its conclusion, an invoice is sent for final payment. Should a dispute arise, however, the case will be taken to arbitration for settlement.

'Cost plus' is more normal for large, long-drawn-out documentation jobs, particularly where government contracts are involved. A retrospective calculation of total costs is made by the representative of the author and an agreed profit margin (usually 7 to 10 per cent) is added to this. It is common in these situations to invoice a client monthly or quarterly.

A limit of liability (or upper ceiling to costs) is often set when accurate forward costing is not possible, or when a client wishes to assess an author's ability without committing himself to the entire operation. This figure may be made available at intervals during a project, and by so doing, the client retains control over the execution and finance of the job.

As an example of the kind of costs which have to be taken into consideration by a product support company producing documentation for a manufacturer, the following guide is given:

Labour –	authors
	illustrators
	tracers
	clerks
	typists
Materials –	paper
	secondary master material
	dyelines
	photocopies
	bromides
Office overheads –	capital equipment
	staff
	premises
	insurance
	expenses

| CHAPTER | PAGES OF TEXT | ILLUSTRATIONS | | | AUTHOR HOURS | | ILLUSTRATOR HOURS | TRACER HOURS | TYPIST HOURS | REPROD | | MISCELLANEOUS |
		TITLE	SIZE	TEXT	ILLUSTRATION				A4	A3	
TOTAL HRS												
RATE PER HR PRICE PER COPY												
COSTS												

TOTAL LABOUR & MATERIAL COSTS.................

Figure 2.5. Example cost estimating form

The cost of the book will vary considerably according to how much weight is given to each of these factors and how they are affected by inflation. Generally, office overheads are thought to work out at approximately 125 per cent of basic labour. The cost of the manual will be labour + office overheads + materials. And the final quotation charge will be the cost + profit + VAT. *Figure 2.5* shows a typical cost estimating form.

Summary

Requirement

Identification of requirement: The following points identify the key areas in eliciting a customer's requirements:

- Specification
- Target readership
- Deadline
- Information available
- Maintenance philosophy
- Subcontractors involved
- Commercial parts incorporated
- Author's contacts
- Validation procedure
- Editorial procedure
- Documentation meetings

Specification

Specification is a document prepared by an authority as a basis for the production of technical literature. In addition we use this term to cover the format, length and presentation of the document. As well as British Standards, there are many military and civil specs, including MOD's Joint Services Publications (JSPs).

Outline design

The outline design anticipates:

- The information the author will need
- The synopsis of the document
- The cost estimate of the work to be done

It summarises both the specification and the requirement, and looks forward to the sources and type of data that will be needed.

Sources of information

Types of information: Information may come in three ways:

- Printed information
- Verbal information
- Visual information

The British Library: The national library of Great Britain which has a number of branches useful to the technical researcher.

Classification system: A method – numeric or alpha-numeric – of allotting to publications a specific classification which distinguishes them from others.

Information search sequence: The following step by step sequence outlines a logical procedure for researching information at a library:

1 Establish subject areas and terminology.
2 Extract classifications from library catalogues.
3 Search index and 'Books in Print' for relevant publications.
4 Order publications not available at library.

5   Compile bibliography of the subject for further research.
6   Consult periodical indexes for state of the art data.
7   Obtain list of trade papers and journals from press guides.
8   Make list of relevant research organisations.
9   Examine yearbooks for appropriate standards and specs.
10 Look through Patent Indexes for useful information.

Universal Decimal classification: A ten-point classification system of which class 6 refers to Applied Science and Technology, and 62 covers Engineering Sciences.

Dewey decimal system: A similar decimal classification, in which class 600 embraces Applied Sciences.

Library of Congress classification: An alpha-numeric method used to classify the publication stock of the US Library of Congress.

International book number: A number prefaced by ISBN and found on the history page of most published books. Used extensively in computerised and teleordering systems.

Contacts

Points for approaching a contact:
- Areas needing clarification
- Queries on modifications
- Technical interpretation
- New information

A visit should accomplish:
- Points answered
- Queries clarified
- Equipment examined
- Way paved for next stage

Meetings

Chairman's main considerations:
- Is meeting necessary?
- Who should come?

- Are they available?
- Subject of discussion?
- What is to be achieved?
- Start time, finish time?
- Venue?
- Advance information?
- Supplementary items?

Agenda: The agenda of a meeting will usually contain
- Place, time and date
- Subject
- Order of discussion
- Other business

Information gathering

Data management: The problems associated with technical information revolve around
- Accumulation
- Retention
- Accessibility

Printed information: May come in the form of
- Previous manuals
- Sales documents
- Specifications
- Test schedules
- Parts catalogues
- Diagrams and drawings

Verbal information is the most difficult to deal with in that it requires use of memory, notetaking, and/or tape recording.

Visual information: All authors should have sight of the equipment they are writing about. This information is best stored in sketch or photographic form.

Copyright

Copyright: An exclusive right allowed by law to an author to print, publish, or sell copies of his original work as he sees fit.

Time limit on copyright: From the time the work is written until 50 years after the calendar year in which the author dies.

The synopsis

Defines the topic breakdown chapter by chapter. It includes the amount of detail to be covered, and the number and type of illustrations to be used. It should comprise:
- Chapter number
- Provisional title of chapter
- Subject of chapter
- Topic breakdown
- Illustrations
- Page estimate
- Other remarks

Costing

Writing terms: Writing tasks are usually undertaken on the following terms:
- Fixed price quotation
- Cost plus
- Limit of liability

Office overheads: Approximately 125 per cent of basic labour.

Cost of manual: Labour + materials + office overheads.

Quotation charge to customer: Cost + profit + VAT.

3

Development phase

Figure 3.1 illustrates the stages of progression through what we shall term the 'development' phase in the production of a technical document. Two loops appear on the flowchart; one for technical 'vetting' (or validation), and the other for editorial scrutiny (editing).

In this chapter we shall examine the steps shown in the flowchart, and also consider the question of 'style' in technical writing.

First draft

With the preliminaries over, the writer can get down to work – the writing of the first draft. This is usually the longest stage of the job, and any self-respecting writer will tell you that he aims to make his first draft the final one – or at least as near to it as possible. Whether this object is fulfilled depends, more often than not, on the attitude of the client. Some will accept a first draft with a few minor adjustments. Others will undergo changes of mind on seeing the typescript, rather like a woman buying a hat. But nevertheless, a good writer will produce a professional product, within tight limits of technical accuracy and editorial acceptability, at the first draft stage.

First draft is essentially about structure. It is *now* that the *actual* shape and form of the finished book materialises.

Some writers will not worry too much about the precise wording at this stage, but will endeavour to cover the subject

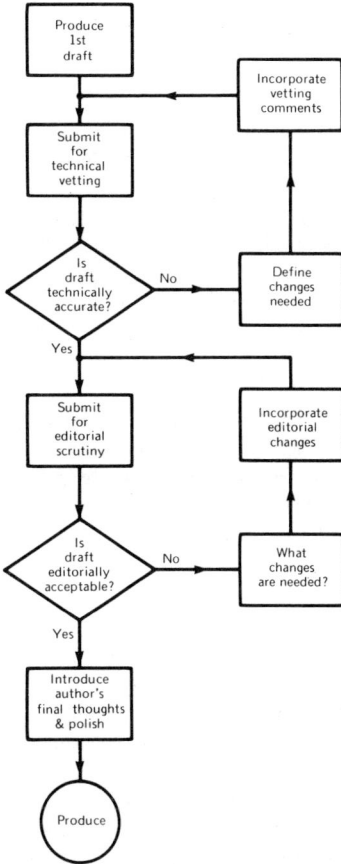

Figure 3.1. Development phase flowchart

in the right depth and with the correct order and emphasis. They then 'tweak' the manuscript into its final, polished state once they have obtained approval for the first, loose version.

While acknowledging the advantages of that approach, my own view is that it is not advisable to show a raw, unpolished version to a client at an early stage simply because first impressions *do* count, and many customers lay great stress – quite rightly – on a good standard of expression in their documentation. It is very easy to lose a customer's confidence by delivering a hastily scrawled, ungrammatical pencil draft, no matter how brilliant it is in terms of structure or technical excellence.

The physical act of sitting down to write is largely an acquired one. It takes a considerable amount of self discipline to tackle the first blank page, with perhaps 300 others stretching out ahead and only a mound of ill-sorted literature and notes as a guide. The prudent author responds to this challenge by breaking the job down into manageable 'bite-sized' chunks.

Initially, he thinks only of the first chapter. Of that chapter he turns his attention solely to the first section – say, three or four pages. In this way he can tackle the work piece by piece, with the end of each section firmly in sight.

Such an approach is more permissible in technical literature, of course, than in the writing of a novel. Technical manuals tend to break down into small sections, and each part, like the product it describes, can be bolted together to form the whole at some subsequent stage of production. The first chapter, for instance, may be a technical description of an item of equipment. Each part of the equipment (hardware part or functional part) can be described in turn, giving the author a series of manageable day-to-day writing blocks that will in time accumulate to form the complete manual. It is rather like building a wall – one brick at a time. Words form pages, and pages amass at a surprisingly rapid rate.

Which brings us to the question: how quickly should an author write? Or in other words, how many pages per day will he be expected to produce? These are imponderables which depend very much on the individual. But in general, it is thought by those who should know, that, taking both the 'learning curve' and the writing together, two A4 pages per day is a fair average.

It is also at this point in the proceedings that the writer comes face to face with the minutiae of the specification.

Heading weights, pagination* (page numbering), and lay-out* of the text and illustrations on the page, must all be incorporated in the draft if it is to have any bearing on the final version. Using JSP 182 as an example, and looking at heading weights in particular, the following shows the permitted range of textual headings in accordance with this standard:

NUMBER ONE HEADING

NUMBER TWO HEADING

1   Ssss sss sssssss ss sss sssssss sss ss ss

Number three heading

1   Ssss ss ssssss ss ssssssssss ssss ssss ss sssss

Number four heading

1   Sss ssss ssssss ss ssss sssssssss ss ss ssss sss

1   Number five heading. Sss sss ss ssssss ss sss sssssss

This type of heading scheme is typical of many technical authorship 'house styles', and repays careful examination. When writing to JSP 182 these weights of heading must be strictly adhered to, for the author will be working for the Ministry of Defence, or an MOD subcontractor. The Ministry demands a high standard of technical authorship from its writers. During 1981 it introduced a new quality assurance scheme to monitor and control the standards of documentation on military projects. *Figure 3.2* Shows an example QA form intended for authors working on MOD documentation.

Pagination methods vary from specification to specification. A simple '1' at lower or upper centre of the page is more often confined to commercially orientated books. JSP 182, by contrast, places both chapter and page numbers in a lower

		QUALITY ASSURANCE PROCEDURE	
		INSTRUCTION TO AUTHOR	

TASK Ref.		JOB No.	
DATE	AUTHOR/CLIENT/LOCAL Ref.		SECURITY CLASSIFICATION

PUBLICATION REFERENCE TITLE

REFERENCE/S OF WORK TO BE UNDERTAKEN

FORMAT	THIS OPERATION (Tick Box)	1st DRAFT ☐
		2nd DRAFT ☐
		CAMERA COPY ☐

TOTAL ESTIMATED TIMES	VETTING COMMENTS INCORPORATED?
Au. III. Tr. Typ.	YES ☐ NO ☐

TO PROJECT LEADER FOR TECHNICAL VETTING	1st SUBMISSION SIGNED	DATE
	2nd SUBMISSION SIGNED	DATE

TO EDITOR FROM PROJECT LEADER FOR EDITORIAL CHECK	1st SUBMISSION SIGNED	DATE
	2nd SUBMISSION SIGNED	DATE

Q.A. APPROVAL BY EDITOR	SIGNED	DATE

Q.A. APPROVAL BY PROJECT LEADER	SIGNED	DATE

COMMENTS

continued overleaf ☐

FINAL MASTER MATERIAL (Camera Copy) ONLY		
FINAL Q.A. APPROVAL BY EDITOR	SIGNED	DATE
FINAL Q.A. APPROVAL BY PROJECT LEADER	SIGNED	DATE

Form No. QA5 Issue 1 Amdt 0

Figure 3.2.   Example quality assurance form for authors

corner of the sheet – bottom left for a 'verso'* (left) page, and bottom right for a 'recto'* (right) page. For example:

| Chap. 10 | or | Chap. 14 |
| Page 12 | | Page 1 |

At this stage, draft illustrations must also be considered, and these are assembled together with the text. Technical illustration is covered in Chapter 5.

Style of writing

There exists a considerable body of opinion which believes that technical English is a 'subset' or dialect of the language, with only a tenuous dependence on 'real' English. The barbarian lobby would prefer to reduce the written language to around 1000 words. This would include technology names and nouns (presumably interchangeable with verbs) strung together with a spare minimum of English words and phrases. Such a language, they argue, would be immune to misinterpretation and could convey technical descriptions in a man *or* machine readable form. It remains a strange phenomenon that technological man, with all his complex artefacts, should look to the future in terms of the palaeolithic past.

However, the grunt brigade have made impressive strides lately among those who believe that technology and its implementation is enough to guarantee human happiness. They have also struck a chord with the stereotyped engineer who is a 'whizz with wires and things' despite his natural illiteracy. The reply to this tendency lies in a statement of the French historian, Renan: 'La Verité consiste dans les nuances' (Truth consists in shades of meaning).

It is probably true that technology and its associated jargon has done more to debase our language than any other influence, including popular culture. The reader of this volume, having considered, or even chosen, technical writing as a career, will possess an interest in the language which is his vehicle, as well as the nuts and bolts of his professional concern. We shall therefore start on the premise that any checks to understanding in technical writing arise from the same sources as those found in other forms of writing. 'Style', we shall assume, is a function of all types of textual narrative, from technical manuals to imaginative fiction.

Good English has certain plain characteristics. It is simple in construction, consistently lucid, and uses a greater number of concrete rather than abstract words. It avoids vague constructions and stale imagery. George Orwell, in one of his essays, names the following faults in modern written English:

Dying metaphors
Verbal false limbs
Pretentious diction
Meaningless words

A dying metaphor is the type of cliché beloved of TV personalities and bureaucrats:

'At the end of the day', meaning, 'when it's all over.'

'I'm over the moon', meaning, 'I'm happy.'

'We shall not be moved', meaning, 'we shall not change our opinions'.

Orwell thought that these prefabricated phrases lost their meaning when used often enough. They strip the language of its natural vividness, and, ultimately, they reduce the awareness of the reader or listener.

'Verbal false limbs' are a form of padding. They come in all shapes and weights, ready-packaged, and portion-controlled:

Render inoperative

Exhibit a tendency to

Having regard to

We are all guilty of this, but civil servants, trade union leaders, and, I regret to say, technical writers, believe they put on a special grandeur when luxuriating in this verbal undergrowth. As Orwell indicates, simple words are eliminated in favour of a glib phrase, usually in the passive voice and with a noun construction (*by examination of* instead of *by examining*).

Pretentious diction includes the use of long Latin or Greek abstractions rather than plain English words:

Expedite the procedure

Ameliorate the situation

This usage has – 'He was conveyed to his place of residence in an intoxicated condition,' when it is simpler to say – 'He was carried home drunk.'

What we are talking about is *jargon*: stale words and phrases whose overt function is to confer an identity on the perpetrator, while exluding interlopers from the magic circle.

Meaningless words could take in all the previous categories. When a technical author writes – for the fourth time in two pages – 'A *caters* for B *enabling* C to *provide* D *ensuring* . . .' he is just going through the motions, semi-comatose, approaching meaninglessness. Words which lack vigour are written by robots, not people.

The hallmarks of a good style are:

- Clarity
- Cadence (balance)
- Appropriateness

The worst enemies of a good style are many, but include:

- Jargon
- Cliché
- Inappropriate ornament
- Somnambulism

Sir Arthur Quiller-Couch, writing in 1915, described 'jargon' thus: 'Caution is its father: the instinct to save everything and especially trouble: its mother, Indolence. It looks precise, but is not. It is, in these times, *safe*: a thousand men have said it before and not one to your knowledge had been persecuted for it. And so, like respectability in Chicago, Jargon stalks unchecked in our midst. It is becoming the language of Parliament: it has become the medium through which Boards of Governments, County Councils, Syndicates, Committees, Commercial Firms, express the processes as well as the conclusions of their thought and so voice the reason of their being.'

Taken to its extreme, bad style is represented by the Indian gentlemen striving for a true English idiom. On the death of his mother he sends the following telegram – 'Regret to inform you, the hand that rocked the cradle has kicked the bucket.'

In contrast, and to sum up this section, let us look at Sir Arthur's peerless definition of 'style': 'This then is style. As technically manifested in literature it is the power to touch with ease, grace, precision, any note in the gamut of human thought or emotion.

'But essentially it resembles good manners. It comes of endeavouring to understand others, of thinking for them rather than for yourself – of thinking, that is, with the heart as well as the head. It gives rather than receives; it is nobly careless of thanks or applause, not being fed by these but rather sustained and continually refreshed by an inward loyalty to the best. Yet, like 'character' it has an altar within; to that retires for counsel, from that fetches its illumination, to ray outwards. Cultivate, Gentlemen, the habit of withdrawing to be advised by the best. So, says Fénelon, ''You will find yourself infinitely quieter, your words will be fewer and more effectual; and while you make less ado, what you do will be more profitable.'' '

Technical vetting

Unless an author is a specialist on the equipment he is describing, his first draft material is more than likely to contain a number of technical errors. This is not so surprising when one considers that an author will spend perhaps two or three weeks absorbing the information that design staff have been studying for years. Despite this, engineers can still become very testy if a writer goes slightly astray on a technical aspect of their machine.

In any well planned project, experienced personnel will establish efficient lines of communication to handle the validation of draft documentation. A quick turn-round will be guaranteed, and no valuable author-time will be lost as a consequence.

Unlike editing, technical vetting is not something a writer will ever be called upon to undertake in normal circumstances. It is the period in every job when the manuscript is removed, and subsequently returns covered in comments in a strange hand, and with whole paragraphs summarily deleted.

Occasionally, when information that had been received from an engineer is removed or changed, the author may suspect that it was wrong in the first place. This is one of the hazards of authorship. Project information changes from day to day. And designers would not be human if they did not seize upon the opportunity provided by the vetting session to tidy up their own thoughts and mistakes.

It sometimes happens too that engineers working on a project fail to grasp certain aspects of their design until they see it described in cold print by a technical author. It is not unknown for them to scamper back to the 'beast' for a swift re-jig, leaving the author's manuscript untended for days in the Quiet Room. After a decent interval the unfortunate writer telephones the engineer to discover what has happened to it. He is informed stiffly that 'certain design changes are under way, which will probably take some time', and would he mind rewriting whole sections of the manual.

This is the essence of frustration. It is also the nature of technical authorship. But despite one's reservations, technical validation cannot be avoided. The process certifies the work's veracity.

Editing

In technical writing an editor is concerned with three aspects of a draft:

- Does it conform to spec or house style?
- Is the flow of material logical?
- Is the grammar and punctuation correct?

Editing is a vital function in the preparation of any document. It is particularly applicable to technical literature because of the need for accuracy and precision. A reader can be misled by badly worded sentences just as surely as by technical inaccuracies in the text.

Normally, a first draft will be submitted for editing once the technical changes arising from validation have been incorporated into the manuscript. It is advisable to present a *typed* draft to the editor, since any document reads better in print than in pencil draft. The editor may be a colleague in the author's office, or he may be a specially appointed staff member in the customer's company. Frequently, the person in charge of the project in the customer's firm will also perform the editing function. He will be working on the widely believed, but false, principle that anyone can be an editor. In this situation, the writer would do well to have his work checked by a fellow author before submission.

It is important, therefore, that all authors are aware of the fundamental principles of editing: not only that they may improve their own work, but in case they are called upon to edit another's.

What then does an editor do? Here is a simple checklist of points to watch for in the editing of any draft manuscript:

- Conformity to specification headings
- Page numbering
- Paragraph numbering
- Conformity of contents list to text
- Layout of illustrations as per spec
- Unusual words or phrases
- Unexplained abbreviations
- Long words that may not be understood
- Unnecessary words – adjectives, adverbs especially
- Undefined technical terms
- Balance of text – length of paragraphs, etc.
- Ambiguities in flow of text
- Punctuation
- Grammatical construction
- Spelling
- Conformity of meaning of text to what was intended

Editing is not an easy job. It is somewhat tedious, and as is apparent from the checklist, requires many literary-type skills as well as a familiarity with the technical background of the subject matter. This is illustrated in the following examples of typical editorial decisions:

'The Large Local Exchange, like all local exchanges, utilises subsystems specially conceived for large local exchanges (LLEs).'

Suggested amendment
As with all Local Exchanges, the large version (LLE) makes use of subsystems specifically designed for them.

The first sentence could be misleading. It implies that all local exchanges, regardless of size, use subsystems designed for Large Local Exchanges. Of course, this may very well be correct, especially if the LLE were designed first, but it seems unlikely. The probable ambiguity can be sidestepped by adopting the approach of the amended version. In the absence of hard information, this is an editor's way of correcting a possible logical error, while covering himself against an outside chance that the facts as stated may be correct.

Stylistically, the first sentence is a tautology and it reads badly. It uses the words 'local exchange' three times in sixteen words. A question of house style also emerges in that 'local exchange' first begins upper case, descends to lower case, then rises again to initial capitals. Obviously a consistent choice must be made, and this is an editorial decision. The author is just as likely to be wrong either way, but since the form of words gains institutional status as an abbreviation (i.e. LLE) it is probably slightly more coherent to adopt the upper case option.

'The equipment will happily run on either 240V or 110V.'

Suggested amendment

The equipment may be run on 110 or 240V.

Inanimate objects should not be anthropomorphised. Machines cannot be happy!

LINE CIRCUITS

Suggested amendment

In the absence of a definite specification ruling on heading weights, attention should be given to the logic of any headings used. In the above case, it is felt that it would be more appropriate to reverse the order given:

TRANSMISSION EQUIPMENT
Line circuits

Good editing comes with practice. Experience is an essential element in the art. But with an adequate mastery of English and a reasonable technical background, there is no reason why any author should not also be a useful editor.

Final draft

When the first draft has passed through the two loops of the development phase flowchart, it should be both technically impeccable (or as near as possible to it) and editorially beyond reproach. The time has now arrived to assemble the final draft for submission to the typist for camera copy preparation.

The final draft must be perfect in all respects, with full instructions for the typist attached, and all points of the specification observed punctiliously. Any changes demanded by the vetting team and editor should have been incorporated, and the author's final polishing – to *his* satisfaction – attended to. It must be stressed that camera-ready copy is never typed until there is full agreement on the draft by all concerned.

At this point a number of smaller tasks must be tackled. The preliminary pages (prelims) should be prepared. These include contents page, key to abbreviations, definitions, and list of illustrations. Also the end pages, which may comprise a bibliography, acknowledgements, and an index. However, the average technical document may not contain many of these, and it is the house style which will dictate the final contents.

When preparing copy for the typist, the following points should be observed:

- Use double spacing
- Write only on one side of the paper
- Write distinctly and legibly
- Use black ink
- Number each sheet ('folio' is the technical term)
- Circle any instructions to the typist to avoid their being typed
- Conform strictly to the house style or specification
- Remember that the line spacing on the manuscript will be observed by the typist unless instructed otherwise

It has been estimated that, on average, a typist makes three errors per page. Some are much better than this, others worse. Suffice it to say that it would be a very unusual typescript if it arrived on the author's desk without a single typing mistake. The typescript should be carefully checked for errors, and any corrections indicated on the accompanying carbon copy . . . never mark the camera copy! A large red 'X' placed in the margin of the carbon copy is normally used to draw the typist's attention to a correction in the adjacent line.

Commercial books

If the manuscript is a book to be published commercially, the typescript submitted will not be camera copy as the publisher will handle the typesetting part for himself. However, this does not remove the writer's responsibility to provide the publisher with a suitably prepared copy. In this section we will look at various aspects of preparing a typscript for a commercial publisher.

All publishers have a house style. Some go so far as to distribute printed bookets to authors setting this out in detail. If an author is writing a commissioned book, or even if he intends it for a specific publisher, he should make an attempt to produce his manuscript in accordance with the required specification. Not only will this save the publisher time and effort, but it will also save money.

In the absence of a style guide, however, a similar volume from the same imprint will provide the necessary information in matters of headings, abbreviations, chapter construction and so forth. It is worth taking some trouble over this, especially if the book is a speculative venture. A well produced manuscript approaching the publisher's house style will be appreciated by those whose job it is to accept or reject it.

It is vital that the pages are typed to a standard pattern – double-spacing between lines and consistent margins. This is to allow the manuscript to be accurately 'cast off'*, or the number of words estimated. Illustrations should be packed in a separate envelope and carefully annotated so that reference can be made to the manuscript sheet where each belongs. Figure numbers should also be clearly marked in the margins of the manuscript at the appropriate place, and a list of illustrations supplied to tie the whole thing together.

The final icing on the cake, so to speak, is the preliminary pages (prelims) and the end pages. These are typed on separate sheets and comprise the following:

Prelims
- *Half-title page* – exhibits the title of the book. No other wording appears on this sheet.
- *Half-title verso* – may carry the author's previous works. The publisher may include any companion or similar volumes from his lists.
- *Title page* – the title, subtitle, if any, author's names and qualifications, if desired, and, if the work has been commissioned, the publisher.
- *Title verso* – the history page, which contains the copyright statement and ISB number, plus a reference to the printer. Eventually, the book's publishing history will appear on this page. A dedication may also be included here, though

it is just as likely to be placed on a separate page following the title verso.

- *Preface* – usually a short piece explaining how, and why, the book came to be written, together with acknowledgements to an army of friends, helpers and counsellers. It is customary at this point to reveal the book's many shortcomings, manfully shouldering all responsibility, and letting the once indispensable army of friends, helpers and counsellors off the hook. A very civilised and benedictory exercise.
- *Contents list* – may contain simple chapter headings and page numbers. In addition, subheadings may be grouped under each chapter title giving full reference to the contents.
- *List of illustrations* – varies considerably in scope. Some books with many in-text illustrations omit this feature. The best plan is to follow the example of a similar book from the publisher's stock.

End pages
- *Conclusions/postscript* – often used to summarise the main message of the book, and to express the author's personal predictions of the future.
- *Glossary* – unusual or technical terms appearing in the text may be amplified here. A good glossary is an indispensable part of an educational book, though sadly frequently omitted.
- *Notes/references* – are better, and more cheaply, gathered together in a separate section at the end of the book.
- *Appendices/annexes* – providing additional detailed material which may halt the flow of the text or is not appropriate there.
- *Bibliography* – a list of books and periodicals pertinent to the subject, and indicating areas of further study for the reader.
- *Index* – vital in a technical book, but often lacking in quality and coherence. The Society of Indexers provide a list of experienced people who will undertake this task for authors, whose responsibilty it remains. In my view an

author should index his own book since only he knows his intentions and the subject matter in full. There are various books on this subject, some of which are included in the bibliography.

The typescript itself may be bound in a multitude of ways. There are many types of elaborate folder or binder on the market. But ease of handling should be the primary concern. David St John Thomas of David & Charles, the publishers, recommends that each chapter be held together by nothing more complicated than a large paper clip! The whole manuscript may then be delivered in the box in which the paper was bought.

Summary

First draft

Learning curve: Used in Tech Pubs to mean the time needed to assimilate the technicalities of a particular project. It is derived from a graphical curve illustrating the rate of learning against time. The curve tends to rise steeply in the beginning before reaching a plateau. If the effort is continued, another jump occurs prior to settling down to a further plateau.

Heading weight: The importance given to a heading by means of its case, position on the page, underlining, italics, etc. A chapter heading has a higher 'weight' than a subheading.

Pagination: Page numbering. Many systems exist, but generally, those used in Tech Pubs will be more informative (i.e. complex) than the ones used in commercial publications.

Style of writing

Man-machine languages: Abbreviated languages which, in theory, can be read both by automated machines and their human operators. Much vaunted in recent years but experiencing problems in implementation.

Style: In the sense used here, style is a function of textual narrative. Its characteristics are, clarity, cadence and appropriateness. Bad style employs jargon, cliché, inappropriate ornament, and a robotic use of stale imagery and outworn phrases.

Technical vetting

Validation: The act of vetting a draft for technical accuracy and balance of interpretation.

Editing

Editing function: In technical writing an editor is concerned with three aspects of a draft.
- Does it conform to spec or house style?
- Is the flow of material logical?
- Is the grammar and punctuation correct?

Principles of editing: An editor checks a manuscript for
- Conformity of headings to spec
- Pagination
- Para numbering
- Contents list reflecting text
- Layout of illustrations
- Unusual word usage
- Undefined abbreviations
- Unnecessary words
- Unexplained technical terminology
- General balance
- Ambiguities
- Punctuation
- Construction
- Spelling

Final draft

Copy preparation: The following points should be observed when preparing copy for the typist:

- Use double spacing
- Write on one side of the paper
- Write legibly
- Number each sheet consecutively
- Circle typing instructions
- Conform to house style

Cast off: Estimating the number of words in a manuscript and its eventual length in terms of pages.

Prelims: Preliminary pages. These may comprise

- Half-title page
- Half-title verso
- Title page
- Title verso
- Preface/foreword
- Contents list
- List of illustrations

End pages: The final pages in a book. These could be –

- Conclusions
- Postscript
- Glossary
- Notes/references
- Appendices/annexes
- Bibliography
- Index

4

Production phase

The flowchart in *Figure 4.1* represents the steps, or series of events, in the 'production' phase of a technical publication. In this chapter we shall consider first certain aspects of camera copy and its preparation. We shall then look at the printer's use of the camera copy, and the author's own task of proof-reading. A discussion of printing methods concludes the production phase.

Camera copy

'Copy' is a term used by writers and printers to indicate the draft of any text which is to be printed. Most of the printing of technical literature nowadays in done by a process called offset lithography* which employs a photographic method of plate-making (see the section on printing). The final copy to be submitted to the printer, or prepared by him, for the photographic process, is consequently known as camera, or camera-ready, copy.

Camera copy is produced by various means depending on the desired quality of the finished product. The most common method for short-run technical manuals and documents uses an IBM 'Golfball' typewriter, which has interchangeable heads to accommodate a wide selection of typefaces.

For an improved standard of finish an IBM Electronic Composer is used. One of the advantages of these machines is that characters are proportionally spaced; in simple terms

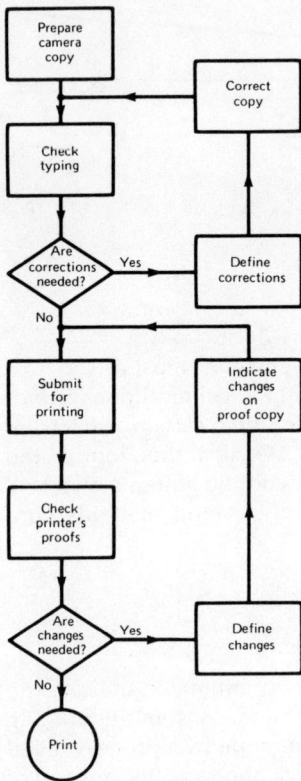

Figure 4.1. Production phase flowchart

an 'i', for example, occupies considerably less width than an 'm'. Although this may seem an obvious and straightforward feature, it is not true of an ordinary typewriter, nor even of an electric Golfball. The Composer is also able to accommodate different type sizes (6 point* to 12 point) and a range of type styles and weights, allowing a finish which approaches the standard set by a typographer.

Another machine used for preparing high quality camera copy is the purpose-built Varityper.

Camera copy may be prepared as a 'paste-up'* in which typewritten, or printed, text and/or illustrations are pasted on

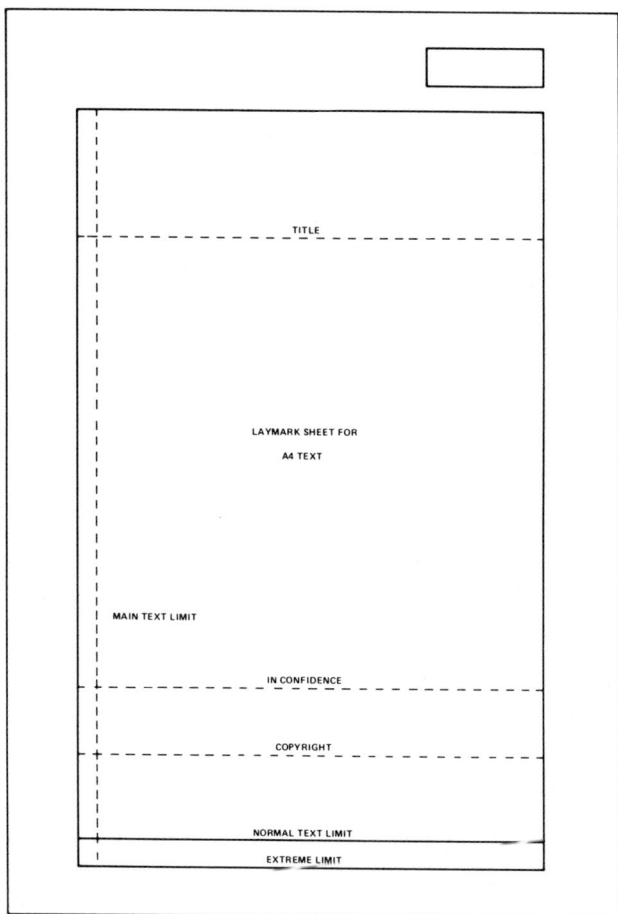

Figure 4.2. Example of A4 portrait laymark sheet

to a separate sheet of paper ready for photography; or it may take the form of directly typed, or composed, text on special 'laymark'* sheets. Laymarks are guide lines, usually in non-reproducible sky-blue, printed on a quality gloss paper, which indicate the margins (sometimes different for verso and recto pages) on the sheet, together with any specified datum lines, such as heading positions or classification notices. *Figure 4.2* is an example of an A4 portrait laymark sheet.

Proof-reading

Proof-reading is done by the author on special sheets provided by the printer. 'Proofs' are example pages taken from the printing medium, as set or photographed, and fall generally into three kinds:

Galley proofs
Page proofs
Machine proofs

The first two categories of proof are representative of the textual arrangement of the type, but not of the final print 'finish'. The machine proof, on the other hand, approximates the condition of the printed book in areas of quality.

Galley proofs are taken at an early point in the typesetting procedure and do not indicate the eventual configuration of the pages in the book. The compositor sets the type in a clamped frame called a 'galley'. The printed sheets derived from the galleys are sent to the author for proof-reading; the aim being to reduce the number of changes needed at a later, and more costly, stage.

The page proofs, indicating actual book pages, are taken when the lengths of run-on text are composed into the correct page sizes. Pagination and other page references can now be incorporated in the text, and the index put into its final shape. Modern books, printed by lithographic techniques, are often proof-read only at this stage.

To obtain an impression of the finished print quality, a machine proof must be requested from the printer. The reason for this is that page proofs are usually taken on a small printing press specially set aside for the purpose. By its nature it lacks the sophisticated facilities of the more elaborate production machines, and does not therefore produce anything like the eventual print finish that may be expected.

Table 4.1. Classified list of marks

General Marks

Instruction	Textual	Marginal mark
Leave unchanged	– – – – – –	
Correct damaged letter/push down risen spacer	Encircle blemish	⊥
Remove extraneous marks, e.g. film or paper edges visible on bromide/change damaged characters	Encircle blemish	✕
Refer to appropriate authority accuracy doubtful	Encircle word(s) affected	

Deletion, Insertion & Substitution

Delete	/ through character(s) or ⊢———⊣ through words to be deleted	
Delete and close-up	/ through character or characters ⊢———⊣	

Table 4.1. Classified list of marks (Continued)

Instruction	Textual	Marginal mark
Insert in text the matter indicated in the margin	⅄	New matter followed by ⅄
Substitute character or substitute part of one or more word(s)	/ through character or ⊢————⊣ through word(s)	New character New word(s)
Wrong fount Replace by character(s) of correct fount	Encircle character(s) to be changed	⊗
Change to Capital Letters	≡≡≡ under character(s) to be changed	≡
Change CAPS to l.c.	Encircle character(s) to be changed	≢
Change to italic	——— under character(s) to be changed	⊔⊔
Change italic to upright type	Encircle Character(s) to be changed	↰
Change to bold type	∿∿∿ under characters to be changed	∿

98

Table 4.1. Classified list of marks (Continued)

Instruction	Textual	Marginal mark
Invert type	Encircle character to be changed	↺
Substitute or insert character in superior position	/ through character or ∧ where required	⌐ e.g. ⌐2

Substitute or insert.

FULL STOP or DECIMAL POINT		⊙
COLON		⊙⊙
SEMICOLON	/ through character or ∧ where required	;
COMMA		,
APOSTROPHE		⌐
QUOTATION MARKS		⌐ ⌐
ELLIPSIS		•••
HYPHEN		⊢−⊣

Positioning and Spacing

REDUCE
    space between
        words
INCREASE
	⋂
	⋃
		⋂
		⋃

Table 4.1. Classified list of marks (Continued)

Instruction	Textual	Marginal mark
Start new para.		
Run on (no new para.)		
Transpose characters or words		
Centre	⌈enclosing matter to be centred⌋	[ ]
Indent	enclosing the text	
		amount indicated 1 em (⅙ in. 12 point)
Cancel Indent	←⌐	
Move matter specified to LEFT	←⌐ enclosing matter to be moved to left	
to RIGHT	⌐ enclosing matter to be moved to right →|	

Table 4.1. Classified list of marks (Continued)

Instruction	Textual	Marginal mark
Raise matter	over ↑ under ⌐matter⌐	
Lower matter	over ⌐matter⌐ under ↓	
Insert space between lines or paragraph	or	

A machine proof is one that is taken from the actual production press and, while it gives a high quality sample, it is an expensive operation, since the machine is temporarily taken out of its high productivity role.

On the question of the costs involved at the proofing stage, authors should know that there is a limit to the number and extent of changes they can make in this area. Generally, printers acknowledge two types of correction:

Compositor's errors
Author's corrections

The first are redressed free of charge – from the author's point of view. Author's corrections, however, are usually limited to a figure around 10 per cent of the total cost of composition. Printers expect – and with some justification – that most deviations in grammar and spelling will have been filtered out by draft copy stage.

A margin is allowed of course, but it should be borne in mind that 10 per cent of costs does not mean that 10 per cent of the text can be amended. The actual figure is disproportionately smaller than this, and any over-run is charged to the writer's account. Indeed, the margin is so tight that there is, for practical purposes, scarcely any room for rewriting the text at this point. An inexperienced author would do well to cultivate a measure of self-discipline in these matters.

The symbols used for marking-up copy for composition and for correcting printer's proofs are shown in Table 4.1, which is reproduced by permission of the British Standards Institution, 2 Park Street, London W1A 2BS, from whom copies of the complete Standard may be obtained.

Each correction is indicated by specific marks, in the margin, and in the text.

Printing

Modern technical books are mostly printed by the lithographic method. However, this technique was not always the market leader. Many printing processes have been invented over the eleven centuries since the Chinese produced the first printed book. All of them fall into one of three categories:

- Relief
- Intaglio
- Planographic

Relief printing, as its name implies, depends on the face of each character protruding above the surface of a small block of alloy into which it is cast. Ink is applied to the face of the letter and adheres to the paper during contact. Letterpress, the oldest method of printing, falls into this group. Originally, type was taken from an 'upper case'* for capitals, and a 'Lower case'* for small letters, and arranged by a compositor in a composing stick, after which the composed line was transferred to a tray known as a galley. Mechanisation brought the Monotype machine, which casts single characters, and the Linotype machine, which casts a whole line of type from molten metal. The galleys are subsequently assembled into pages, a number of which are locked together in a steel frame called a chase, making up a forme, from which one side of a sheet of paper forming a book section may be printed.

The letterpress process is flexible and reliable, if laborious, and maintained its pre-eminence for five centuries until photographic lithography replaced it in many applications.

Intaglio printing is the opposite of the relief method. Examples are gravure (line engraving) and photogravure (a photographic process using dots of varying size to form the image). In both variants, the characters are cut or etched *into* the surface instead of standing out above it. Low viscosity ink is applied to the surface so that the etched gullies are filled. The remaining ink is mopped or scraped away. When paper is applied to the printing plate, the ink in the gullies is lifted out by suction.

In the *planographic process* the image is situated on the surface of the plate. The principle underlying the process is simplicity itself: grease repels water but retains ink. The characters are first impressed on the plate, usually by photographic means, and coated with a thin layer of grease. Water is washed over the surface, but is rejected by the greasy areas. On application of the ink, the grease on the image retains it, while the water over the rest of the plate effectively repels it.

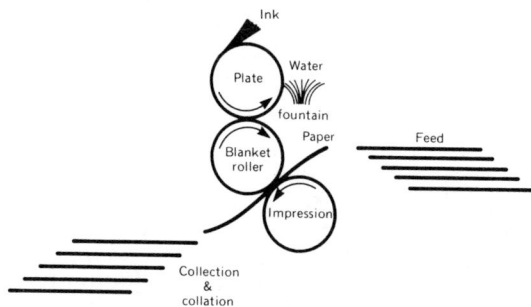

Figure 4.3. The process of offset lithography

At the beginning of the present century, *offset* lithography was developed in Germany. This improved method involved transferring the matter to be printed from the plate by an intermediate cylinder to the paper. *Figure 4.3* illustrates the process. Offset litho was, at last, a cheap, flexible method for the printing of large quantities of paper. And combined with the fact that the plates could be produced ever more quickly by chemical transfer, direct photography, or electrostatic processes, the scene was set for a strong movement towards this method.

Somewhere between intaglio and planographic processes come silk-screen printing and wax-stencil duplicating, where the ink is squeezed through a master from the back to the front onto a flat surface. Silk-screen printing can be used on a variety of surfaces from cloth to metal producing a high-quality image, whereas wax-stencil is used primarily for low-quality, low-cost duplicating.

Colour printing is effected by making a series of passes over the same paper, each in a different colour. The combination of primary colours, if correctly aligned, produces a fully chromatic image.

An author should have a basic understanding of these methods and, perhaps even more so, of the main areas in which each is used.

Photogravure produces high quality colour reproductions, but is expensive to the point of being uneconomic for print runs of less than 6000 copies. Quality, however, remains high for up to 20 000 impressions, and it has the characteristic of producing solid areas of colour, which are not broken up by white gravure lines. Photogravure is often used for magazines, and some high quality work, but it is never used for handbooks or technical manuals because of the cost factor.

Letterpress gives constant quality throughout a long run, and has the advantage, in flat-bed presses at least, that alterations to the text can be made without recasting a plate.

Lithography required some attention throughout a print run to maintain quality, but is is suitable for both short and long runs and has the edge over rivals in cost and speed.

Summary

Camera copy

Copy: a term used by writers and printers to indicate the draft of any text which is to be printed.

Production of camera copy: camera copy is usually prepared using

- Golfball typewriter
- Electronic Composer
- Varityper

Proportionally spaced: a phrase used when letters are allowed a space on the page which is proportional to the actual width of the character. The type used on an ordinary typewriter is constant and therefore does not exhibit proportional spacing.

Preparation of camera copy: may be pasted-up by sticking printed, typed or composed text and/or illustrations onto special sheets of the correct size ready for photography. An alternative is to type the material directly onto laymark sheets, which contain pre-printed datum lines to guide the typist.

Proof-reading

Types of proof: There are essentially three types of printer's proof

- Galley proofs
- Page proofs
- Machine proofs

There is an intermediate stage of proofing referred to as 'page-on-galley' proofs. The first two categories indicate the textual arrangement of the type, but not of the final print finish. The machine proof approximates the condition of the printed book in areas of quality.

Corrections made at proof stage: two types of correction are acknowledged by printers

- Compositor's errors corrected by printer or author.
- Author's corrections indicated during proof-reading.

The first are usually made gratis, but the second may be limited to 10 per cent of the total cost of composition.

Printing

Categories of printing methods: three groups may be identified

- Relief printing
- Intaglio printing
- Planographic printing

Relief printing depends on a raised character transferring ink directly to the applied paper. *Intaglio* methods use cut gullies to form reservoirs for the ink, which is then lifted off onto the paper by suction. *Planographic* techniques hold the image on the surface of the printing plate, utilising the principle that grease repels water. *Offset lithography* involves transferring the image from the plate to the paper by means of an intermediate roller.

Uses of printing methods: Photogravure is uneconomic at print runs of less than 6000. It is used for high quantity colour work such as magazines and some art prints. Letterpress is useful for long runs because it maintains its quality throughout. But it has lost ground to the cheaper and more compact lithographic methods, which are excellent for both short and long runs. New filmsetting methods allow litho prints to match the quality of the best letterpress.

5

Technical illustrations

By the very nature of things a technical writer is rarely a competent illustrator. The converse is also true, but perhaps less so: some illustrators do engage in a truncated form of writing when asked to provide lengthy captions for illustrations, or even to write a skeleton text linking their own artwork. Generally, however, the 'horses for courses' principle applies and, in the course of producing a technical book, author will liaise closely with illustrator, and vice versa.

So, the technical writer should understand the fundamental features of technical artwork, certainly to the point of being able to communicate in the terminology.

In this chapter we shall briefly consider the products of a technical illustrator: diagrams*, line illustrations*, and halftones. It is recommended that prospective authors devote some time to this topic. A companion volume in this series, 'Beginner's Guide to Technical Illustration' will repay study. In the meantime, this chapter will serve as a brief introduction to the subject.

Diagrams

Diagrams proliferate in the work of a technical writer. They are probably the commonest form of information presentation. Diagrams are two-dimensional line representations, usually intended for line block or line plate reproduction. They often employ symbols or simplified blocks to represent the objects or functions involved. These symbols may be the subject of standards or specifications, with strictly formalised sizes and shapes. Many symbols can be bought ready-printed as rub-on transfers allowing some economy on illustrator-time. Symbolic diagrams include graphs, maps, charts, hardware and functional drawings, circuit diagrams and flow-charts.

Diagrams are useful in the presentation of statistical, symbolic or functional information. But more elaborate forms are needed for an adequate depiction of engineering hardware.

Line illustrations

The most common illustrations for equipment in technical books are line perspective drawings. These pictorial drawings are designed to convey the shape of objects and the position in space of the relevant parts. They may also illustrate the components used in assemblies and sub-assemblies, and the way they fit together and come apart.

Line illustrations are reproduced in a single solid colour with no tonal distinction. They may be black and white or coloured, using mechanical tint*, stippling or hatching. Subjects printed by line block or line plates have no intermediate grey tones, and consist of entirely of lines, dots or solids. Tones may be simulated by various means, usually by hatching, or by rub-on 'tints', but the lines must be sufficiently far apart to reproduce on a line block or plate.

Most illustrations are produced 'twice up'* on Bristol board (a smooth, double-surfaced white board available in various thicknesses), or on linen, by tracing over a pencil draft. It can be an expensive business, mainly because two drafts are necessary: first a pencil version, then an inked-in final copy. But the term 'line illustration' does not only apply to pen and ink drawings. Technically, it can be used for any illustrative material which has no tonal variation, and which is therefore suitable for line block or line plate printing.

The pencil draft consists of the construction lines used in laying out drawings for inking-up. Sometimes a directly traced version is possible from other illustrations. A 2H or harder pencil lead is used for this work, and the aim is to produce very fine lines, which will erase neatly from the final copy without smudging.

Inking-in is performed with two basic types of pen:

Drawing pen with adjustable nib.
Rapidograph pen.

The standard drawing pen is supplied with a calibrated nib arrangement, which may be varied according to the line thickness desired. Rapidograph pens come in a range of nib thicknesses from, 0.1 mm to 1.2 mm.

Two types of ink are used in this work:

Indian drawing ink (jet black).
Plastic ink.

Indian ink us used for drawing on linen, but, because of its flaking characteristics, it is not suitable for tracing on plastic or film.

Other equipment used in the production of line illustrations includes stencils for drawing in standard symbols, adhesive symbols which are stuck directly onto the drawing, instant lettering in the form of rub-on transfers, and mechanical tints which may be cut to size and rubbed onto the required area in order to simulate tonal gradation.

Realism is added to line illustrations by means of perspective, which we shall now consider.

Perspective drawings

'Perspective'* is an artist's attempt to come to terms with the world as its exists. In order to represent objects in space as they really are – or seem to be – on the flat plane of a sheet of paper, various expedients have to be employed. The essence of the 'trick' is to *project* the three-dimensional object onto the two-dimensional surface in such a way that the apparent linear relationships of receding planes are maintained, or, in some cases, exaggerated. There are three types of linear perspective:

One-point (parallel)
Two-point (angular)
Three-point (oblique)

Figure 5.1 shows these differences in terms of three cubes. Distortions would appear if, for example, the one-point drawing showed two faces of the cube, or the two-point view illustrated three faces. Only three-point perspective allows a true depiction of three faces of an object. For this reason it is usually employed in technical illustration.

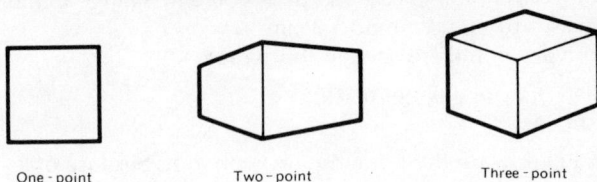

One - point   Two - point   Three - point

Figure 5.1.   Three types of linear perspective

Photographs may sometimes be used for line drawings as an example or guide for the illustrator. This is not as simple as it sounds, however, since there are many problems associated with camera angle and obtaining the right axis or axes through the equipment. Photographic prints may be useful as a simple introductory guide for exploded views or cutaways.

A further technique is to draw in the outlines and relevant details on a photographic print with ink, and to bleach out the image with potassium ferricyanide leaving the line drawing intact. This is not frequently used in technical work nowadays, but it can be cheap and effective in certain circumstances.

Half-tones

The half-tone process is used to reproduce any subject with continuous varying tones, such as photographs, shaded-in or wash drawings, air-brush work, and drawings in which the

110

lines are so closely spaced that they would reproduce as tonal gradations.

Half-tone illustrations are generally more costly to reproduce than line drawings. Photographs may need retouching by expert hands, or the tonal contrasts may require heightening. Another consideration is the compatibility of the printed half-tone with the rest of the artwork, especially since it demands a certain quality of paper for satisfactory reproduction. Line drawings may sometimes be more appropriate, and cheaper, for the subject matter or presentation required.

In order to reproduce a continuously toned subject, a printer overlays the photographically sensitive material with a screened (grid-like) glass plate or negative, which has the effect of breaking up the image into a pattern of dots. Dark areas of the original are reproduced by large dots with less separation than the smaller points representing lighter areas. In this way, tone and shape are built up on the letterpress block or litho plate.

On a letterpress block, dots are raised to the same height as the character type. On litho half-tone plates, the dot pattern is imprinted on the surface.

Validating technical illustrations

Technical illustrations need to be checked in much the same way as the accompanying text. In complex drawings, authors should watch particularly for omissions of lines or annotations. Annotations may be direct or indirect. That is, typed on the drawing, or keyed to a separate table. In the latter case all cross-references should be confirmed.

The following checklist suggests points to look for when vetting a technical illustration:

- Has the specification been adhered to?
- Is the format correct?
- Are there any technical inaccuracies?
- Is the titling correctly placed and worded properly?
- If the illustration is to be reduced, is the lettering of sufficient size to be readable?
- Is the page identity or figure number accurate?

- Is the security marking there, and correct?
- Is the layout and composition clear?
- Is the line thickness to spec and uniform?
- Are cross-references correct?
- For double-page spreads, is the material which crosses the fold entirely clear?
- For simulated shadow-lines, are these placed correctly and uniformly?
- If a variety of tints are used, have they been correctly indexed?

The future of illustrations

The future of technical illustrations is perhaps not so clear as it once was. Computer-aided methods are invading almost all technical areas, and none more quickly than that of graphics. Software packages are now available which will perform wonders with engineering drawings; and the continuous-

Figure 5.2. Graphics computer with integrated printer (by courtesy of Hewlett-Packard Ltd)

Figure 5.3. VDU and plotter (by courtesy of Hewlett-Packard Ltd)

perspective, three-dimensional representation is now commonplace. Machines are on the market which produce a 'hard copy' or print-out of remarkable quality from the image built, or called up, on the display screen (*Figure 5.2*).

The future is always difficult to predict. But it seems certain that, apart from purely artistic areas, much of the illustrative work for books will, before very long, be done by engineers or authors at a visual display terminal. Whether authors should welcome this situation is open to speculation. One day perhaps, they themselves may be replaced by creative microprocessors capable of original thought!

Summary

Diagrams

Diagrams are two-dimensional line representations, usually intended for line block or line plate reproduction. They are used in the presentation of symbolic, statistical, and functional information.

Line illustrations

Reproduction of line illustrations: these pictorial drawings are reproduced in a single solid colour with no tonal gradation. They have no intermediate grey tones and consist entirely of lines, dots or solids. Illustrations are usually produced twice up on board or linen.

Pencil draft consists of the construction lines used in laying out drawings for inking-up.

Inking-in: two types of pen are used in this process

- Standard drawing pen
- Rapidograph pen

Inks used are:

- Indian drawing ink (jet black)
- Plastic ink

Perspective drawings: Realism is added to line illustrations by means of perspective

- One-point (parallel)
- Two-point (angular)
- Three-point (oblique)

Three-point perspective is usually incorporated in technical illustration because it is the only form which allows the depiction of three faces of an object without distortion.

Vanishing points are dots placed at the periphery of a perspective drawing to which parallel lines apparently converge. This gives the impression of actual perspective on a two-dimensional plane. In three-point perspective, three vanishing points are used.

Half-tones

Half-tone process: used to reproduce any subject employing continuous tonal gradation. The material is overlayed with a screened negative or glass plate, which breaks up the image into a pattern of dots of apparently varying density.

114

Validating technical illustrations

Checklist for vetting illustrations:

- Has spec been followed?
- Is format right?
- Are there technical errors?
- Is title correct?
- Is lettering of sufficient size?
- Are page and figure numbers correct?
- Is security marking accurate?
- Are layout and composition clear?
- Is line thickness correct?
- Are cross-references correct?
- Is material crossing spine clear?
- Are shadows correctly placed?
- Have tints or colour codes been correctly indexed?

6

Materials and equipment

In the course of his day-to-day work, a writer uses a large variety of materials and a wide range of equipment. In the present chapter we shall examine some of the more important of these:

- Paper
- Typewriters
- Copiers
- Microform
- Word processors

Paper

Nothing is taken more for granted in our throw-away society than paper. From its beginning as a prized and costly commodity, it has achieved the distinction of becoming the most disposable of all items. Yet paper is big business. Whole industries have grown up around it.

Books, of course, depend on paper. And authors use vast quantities of it in the course of their work. But few people know very much about it. Perhaps this is not surprising. There are, for example, no fewer than eight British Standards on various aspects of paper. They are reproduced here for their curiosity value:

BS 4000 1968 Sizes of papers and boards
BS 1413 1970 Paper sizes for books
BS 2489 1971 Sequence of measurements for printed matter
BS 1342 1962 Detail drawing paper

BS 1343 1962 Cartridge drawing paper
BS 1808 Part 1 1970
Part 2 1967 Sizes and layouts of commercial stationery
BS 3429 1961 Sizes of drawing sheets
BS 4264 1967 Envelopes

Paper sizes are a harrowing business. As with most standards derived from mediaeval history, they have a quaint complexity.

Since the early 1970s, British sizes have been largely superseded by the ISO metric system – but not entirely. BS 1413 recommends the use of the A4 and A5 sizes, plus the metric equivalents of eight of the old British sizes. The complications do not end there, though, since there are subtle differences between British and European nomenclature. The British practice was to take a so-called 'untrimmed' (meaning, 'not guillotine-trimmed') sheet of paper, and to trim it in accordance with the printer's or publisher's own idea of the particular size. Thus nominally identical sizes could vary depending on which source performed the processing.

British sizes were – and still are – based on the 'Broadside'. This can be any size of paper supplied by the papermaker. When the Broadside* is successively halved along its longer edge (see *Figure 6.1*) the divisions are called respectively:

Folio
Quarto
Octavo
16mo

These divisions are not paper sizes as such. The actual sizes are defined by quoting the division with respect to the original size of the Broadside sheet. For example, Large Crown Quarto, or Demy Octavo, would be derived from Large Crown and Demy Broadsides respectively. These names come from an age when printing and publishing still had a flavour of mystery and romance.

The ISO sizes are less flamboyantly named, though not less interesting for that. They are based on the ancient classical

117

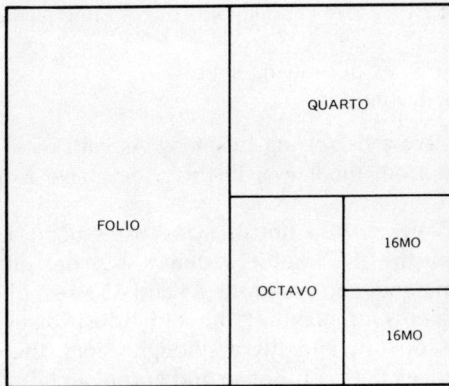

BRITISH PAPER SIZES

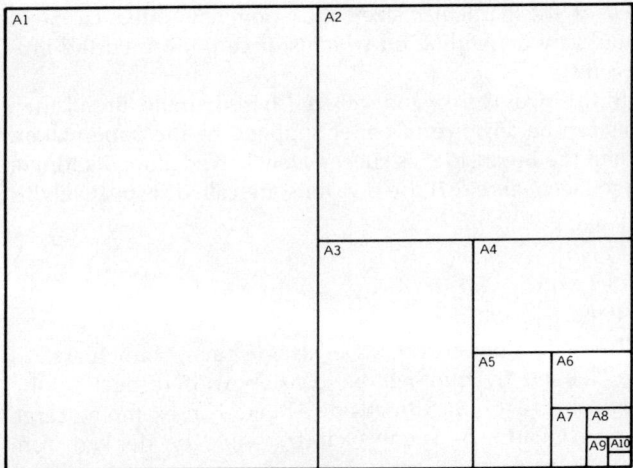

ISO 'A' PAPER SIZES

Figure 6.1. British and international paper sizes

(some would say Hermetic) principle of the 'Golden Mean' or 'Golden Square', much used by architects and artists during the eighteenth century.

The Age of Enlightenment and Reason has given us a paper size in which the ratio of the short side to the long side equals the ratio of the short side to the diagonal of a square based on the length of the short side. For those less mathematically gifted, it follows that $y = \sqrt{2}x$, which is, of course 1.414x, or 1 : 1.414.

Clearly, whichever system you prefer will depend very much on character. British Standards, true to tradition, have effected a compromise, and the following list covers most of the sizes met by technical writers.

	mm	inches
A0	841 × 1189	33.11 × 46.81
A1	594 × 841	23.39 × 33.11
A2	420 × 594	16.54 × 23.39
A3	297 × 420	11.69 × 16.54
A4	210 × 297	8.27 × 11.69

The metric equivalents of the eight traditional British sizes are:

	Quarto		Octavo	
	Trimmed	Untrimmed	Trimmed	Untrimmed
Metric Royal	237 × 312	240 × 318	156 × 234	159 × 240
Metric Demy	219 × 276	222 × 282	138 × 216	141 × 222
Metric Large Crown	201 × 258	204 × 264	129 × 198	132 × 204
Metric Crown	189 × 246	192 × 252	123 × 186	126 × 192

Typewriters

Although most technical authors write their drafts in pencil or ink, and submit them in this form to their typists, there are some, especially freelances and writers of commercial books, who do their own typing. For them, the portable mechanical typewriter is one of the great inventions. But from a technical point of view, it does have its shortcomings, particularly in the preparation of camera copy. The main drawbacks are:

- Only one typeface of one point size is fitted to any one typewriter.
- Because of the different pressures applied to the keys, the density of the typed characters is not consistent.
- Only very rough justification* of the right margin is possible.

The electric typewriter partially overcomes these problems. The IBM Executive, for example, is widely used in technical publications because of its proportional spacing* feature and consistent quality. While the typefaces are not interchangeable on the machine, various models are marketed with different typefaces. Justification is also feasible, but involves some complicated calculations.

The IBM 'Golfball' model is similar to the Executive, except that the characters are carried on a centrally placed interchangeable head, or 'golfball'. As the keys are pressed, the head revolves on a vertical axis until the required character or symbol is adjacent to the paper. It is the head itself which moves across the paper from left to right, in contrast to a mechanical machine in which the carriage moves to the left across the central typing area. A variety of typefaces and sizes are available.

The Varityper is a useful machine which may be used for both unjustified and justified text. It has a choice of typefaces and type sizes which range from 6 to 12 point.

Copiers

The two most popular kinds of copier now used in publications departments are:

Dyeline or 'diazo' copiers.
Plain paper copiers (like Xerox or Nashua).

Dyelines are used mainly for duplicating large engineering drawings and circuit diagrams. They are also used for producing 'secondary masters'*, which are copies of original masters printed on translucent material.

A secondary master is often used when amendments are necessary to an original illustration. Any changes are made

120

on the secondary copy, and further dyelines run off from this. It is a cheaper procedure than a complete rehash of the master material. Secondary masters are prepared on the dyeline machine, but using film instead of normal paper.

The dyeline process is one of the oldest copying methods still in use. The translucent master illustration is placed in contact with sensitised paper and fed into the machine by a turning roller. The paper is exposed to ultra-violet light, and subsequently developed in either an ammonia solution (for the wet process) or vapour (for the dry process).

Plain paper copiers are now used almost exclusively for the mass duplication of most kinds of material. Excellent copies may be obtained using ordinary paper, though the best results are achieved with higher quality glossy paper. These copiers can have many additional features, such as a reduction facility and automatic collating.

Microfilm

With the falling cost of memory devices and their increased packing density, some may feel that microfilm has had its day. It is hard to quarrel with this argument, at least in the long term. But the fact remains that microfilm is still a very popular, and cheap, form of document storage.

Microfilm was developed as a storage medium in response to the general expansion of paperwork in business and large organisations. It reduces huge quantities of paper to the size of a small sheet or roll of film. The information can be 'retrieved' by enlarging the image on the screen of a microfilm reader.

Some publications are now committed wholly to microfilm (micropublishing), an expedient which sidesteps the printing stage. The recipients of these films are spared the weight and space problems associated with bulky documentation, leaving only the handling of 8 or 16mm film cassettes. As a guide, the contents of a 12 volume dictionary could be accommodated on 2 square inches of film, equivalent, in computer terms, of a thousand million bits per square inch (lkMbits/sq in).

Microfilm is a useful back-up medium, and can be used as an insurance by companies with their databases on magnetic tape or Winchester disc. Magnetic media are subject to corruption in a variety of ways, and it would not make sense to keep a complete paper duplicate of an organisation's records.

The advantages of microfilm can be summed up thus:

- Saves time and space, hence cost
- Convenient to use
- Security and other hazards are minimised
- Easy to transport
- Efficient for cataloguing
- High storage capacity
- Easy to amend
- Easy to reproduce

When used for illustrations, microfilm is equally cost-effective. The drawings are photographed on individually mounted 35 mm film, together with the title, references and issue number. Similar advantages are claimed:

- Bulk is reduced by up to 90%
- Easier and cheaper to post or transport
- Amended up-issues may be stored without the need for intermediate masters
- Further copies can be inexpensively produced as an insurance against loss or damage

A fuller list of microform systems follows:

- Aperture cards – usually 35 mm
- Roll film – 35 or 16 mm; 70 or 105 mm
- Jacket fiche – usually 35 mm
- Fiche – usually 98 frames on 16 mm
- Microfiche – 24 to 98 frames on 105 mm film in 148 mm length
- Superfiche – 200 frames
- Ultrafiche – 3000 frames

Word processors

It was only a matter of time before the computer caught up with the efficient author. At present, computers are hovering on the periphery of his consciousness, leaving his creative areas intact, but invading the typing, editing and storing functions in the form of word (or text) processors.

A word processor is a computer, interfaced with a keyboard and a printer. Its operating system is specialised for processing alpha-numeric strings and has a high quality editing function. The cost of these machines is falling so rapidly that it may not be fanciful to imagine a situation, within 5 to 10 years, when many authors will write directly onto a VDU (Visual Display Unit) deleting and correcting as they go. At the push of a button, the day's output will be impeccably printed at high speed – instant camera copy! One speculates, however, whether the novels of Dickens or Dostoievsky would have quite the same atmospheric quality had they been composed on a word processor.

Summary

Paper

British paper sizes are based on the 'Broadside' – a large sheet of paper in varying sizes supplied by the papermaker. Broadsides are now classified into four sizes:

- Metric Royal
- Metric Demy
- Metric Large Crown
- Metric Crown

British Standards recommend two divisions of these Broadsides:

- Quarto
- Octavo

A *Quarto* is a quarter of the Broadside sheet, and an *Octavo* is an eighth.

ISO paper sizes are derived from the 'Golden Mean' of classical times. The ratio of the shorter side of each sheet to the longer is $1:1.414$. Sizes in common use are:

● A0
● A1
● A2
● A3
● A4
● A5

Typewriters

Faults of mechanical typewriters mainly fall into the area of camera copy production:

● Only one typeface and size
● Character density not consistent
● No justification facility

Electric typewriters overcome some of the problems associated with camera copy preparation, including:

● Proportional spacing feature
● Consistent type density
● Justification possible
● Different models carry different typefaces or sizes. Some makes have changeable heads.

Typewriters used for camera copy production include IBM Executive, Composer, Selectric 'Golfball', Varityper.

Copiers

Dyeline copiers are used for duplicating large engineering drawings, circuit diagrams, and so on. Can also produce secondary masters on translucent film. The process relies on a sensitised paper which is exposed to ultra-violet light before development in ammonia.

Plain paper copiers use ordinary paper for copying. Produce good quality image: many copies are possible.

Microfilm

Microform: the whole 'genre' of reduction copying on film materials. Advantages include:

● Time and space saving
● Convenient
● Hazards minimised
● Easy to transport
● Efficient for cataloguing
● High storage capacity
● Easy to amend
● Easy to reproduce

Word processors

Word (or text) processors are computers interfaced with a keyboard and printer.

7

Miscellaneous matters

Technical writing is rather a mixed bag of a career. The qualities required of a writer vary from literary proficiency to capabilities of man management. In between, there are the technical skills and a knowledge of many divergent disciplines. A writer usually requires some aptitude for management in its wider sense, and also the elusive ability to prise information from reluctant sources: engineers, busy technicians or frantic designers behind schedule.

A technical writer may be pictured as the co-ordinating centre of a complex web of trades and professions (*Figure 7.1*). He must exercise his whole range of skills in order to accomplish his task – a book, a manual, or the entire documentation of General Motors.

This chapter examines some of these peripheral aspects of the subject. It is not possible to cover everything here, but the prospective writer will get a glimpse of the kind of concerns he may have to handle. It begins with a section on *translations* – points to look out for when preparing technical material for a translator. A brief consideration of *indexing* is followed by a section on the *Development Documentation System* (DDS), and a look at the documentation systems used for the maintenance of complex modern equipment. A discussion of *network analysis* concludes the chapter.

Translations

The translation of technical material presents both writer and linguist with special problems. In common with imaginative literature, there is no such thing as a perfect translation. Several translators, given identical texts, will produce different solutions to the puzzle. There is, however, a median line through the problems, balancing accuracy of fact with interpretative freedom.

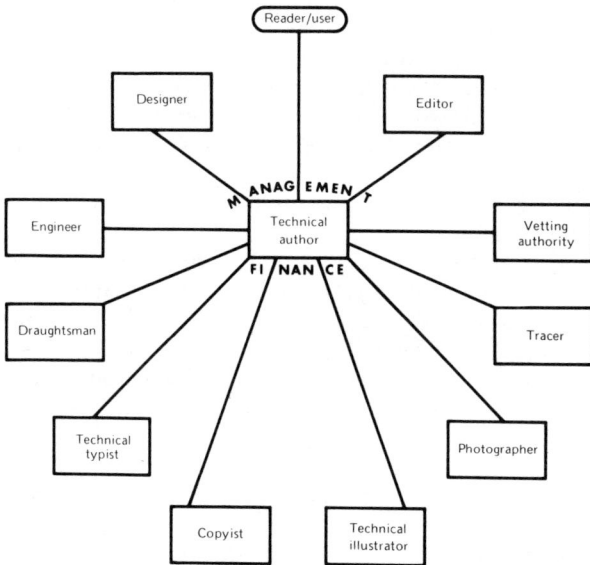

Figure 7.1. A technical author's world

The exact end product will, of course, depend on the requirements of the particular job. There are generally three types of translation:

● A literal translation rendering the original phrase-by-phrase.

● A free translation giving the translator wider scope for adaptation into the local vernacular.

● A partial translation in which a summary is made of the original text.

A snappy sales brochure, for example, requires above all else a true colloquial rendering for the target community. It would obviously be facile trying to sell a household appliance in a strange stuttering text bearing no relation to everyday speech. Either the translator must be a competent copywriter with the freedom to adapt the material to local requirements, or his verbatim translation should be give to a copywriter in the country concerned for further processing.

At the other extreme, a description of a maintenance procedure for an intricate piece of equipment with stringent safety precautions would call for a strictly formal translation, in which each sentence is rendered literally into the new language.

A writer producing a text which he knows is to be translated into another language should bear certain points in mind. A badly edited or poorly punctuated document will cause problems in the original let alone in translation. The writer should always ensure that his meaning is clear and unambiguous, avoiding inconsistent terminology, the use of colloquialisms, and any other loose deviations from a tight and accurate text which a translator will need to do a good job.

Choosing the right translator is an important consideration from the outset. Starting on the premise that a practitioner works best when translating *into* his own native tongue, two other points immediately arise: his understanding of the source language and his technical competence to handle the subject matter. Both aspects could be crucial to the quality of the finished product.

How does an author engage – or even go about finding – a suitable translator? Fortunately the field is a wide one:

● Local contacts – foreign wives or husbands of friends or colleagues, etc.
● Yellow Pages and other directories where translators or bureaux are listed.
● The British Standards Institution can often help through its translating and advisory service.

● Embassies/High Commissions/Consulates who may keep lists of their nationals capable of offering a translating service.

Translating is a well paid occupation – or so it seems from an author's viewpoint! The price of translation, like typing, is normally quoted as a rate per thousand words. For smaller documents, a unit called a 'Folio', or 75 words, is used, and is the least that will be charged for. As in dealings with all independent professionals, though, charges may be negotiable, particularly if any unusual considerations apply.

The main factors likely to put *up* the cost of any translation are, the language to be used (i.e. non-Roman scripts like Arabic, Japanese, or Russian will be significantly more expensive), and any special requirements such as side-by-side translations or additional copywriting skills. Cost assessment will be based on:

● Time factors.
● Technical content.
● Specialised subject matter.
● Specifications or standards to be observed.
● Language – European/Non-European/Non-Roman.
● Any proof-reading or other tasks.

English is now the major world language in terms of international communication. In areas of science and technology the dominance is even more pronounced. Over the years this has tended to give the British a false sense of security. Somewhat disastrously, they have assumed that the rest of the world will buy British goods on British terms – complete with a manual in English, or at best a poorly translated equivalent.

This insular mentality has prevented any great leaps forward, but one way of counteracting it is to make sure that all technical and customer support documentation receives the best possible translation effort in the language of the target population. Efficient communication overcomes many other deficiencies.

Indexing

A good index is essential, even if it is not always provided. Indeed, the reference value of a technical book may be directly proportional to the quality of its index.

Generally, two types of index are used:

A general index at the back of a book – sometimes divided into 'names' and 'subject' indexes.

A detailed contents list (so detailed that it serves the function of an index) with subheadings – perhaps one for each chapter.

In technical handbooks, the contents list breakdown is normally employed. A list of abbreviations or a glossary are occasionally added, but a full index is often omitted completely. The reason for this is partly that the book must be updated at frequent intervals and may never be 'fixed' into any final form. It may also be the case that few copies will be required and the standards (and costs) incurred for a commercial work would be inappropriate.

Commercial technical books of any weight or reputation include a general index, which is normally the last item in the end pages. The index is the responsibility of the writer and will either be compiled by him or placed, at his expense, with a professional indexer. If the book is part of a series, the depth, and hence length, of the index will be laid down by the series house style. Otherwise, an author should aim to make his index as comprehensive and comprehensible as his resources of time and patience allow.

An index can only be finally completed when the book is in page proof. But much of the groundwork can be prepared in advance. In the period between sending the manuscript to the publisher and receiving the proofs, the author should go through his copy of the book, extracting the items of subjects required for the index. These can be written out on cards or strips of paper ready to be put into alphabetical order.

The complexity of the index will depend on the scope of the subject, and the depth of treatment applied. It may be straightforward as in:

Or all-embracing as in Hugh Thomas's *Unfinished History of the World* in which the index takes up 41 pages:

Several aspects may be mentioned here. 'Use of wet nurses, 52, 53;' as against, 'breast and bottle feeding, 53–4.' The difference shows that the wet nurses are referred to only in passing on pages 52 and 53, whereas breast and bottle feeding constitute a substantial section over pages 53–4.

If a number of trivial inclusions occur over a sequence of pages, the device 52ff may be employed.

The reference to 'fall in age of sexual maturity, 232n' indicates, by the use of *n* after the page number, that it is included in a footnote on that page.

If a particular topic occurs more or less continuously throughout a work, the word *passim* can be used instead of page numbers.

Subjects may be cross-referenced if this helps the reader: 'Printing . . . see Lithography'. Too zealous an application of this device, however, can be tedious, not to say confusing.

The Development Documentation System (DDS)

DDS is a documentation procedure/methodology employed in system design. It can be used for both electrical and mechanical engineering disciplines, covering hardware, software, logic and functional dimensions. It has become a versatile tool for designers of any system, allowing detailed recording of progress as the project develops. It is not a *publications* application, but more a methodology and information system for development purposes. Writers, however, may be involved in its implementation, and will refer to it during the writing of other project documentation.

The Development Documentation System was begun in the 1960s by the Naval Applied Science Laboratory in America. Since then much work has been done on it in the UK, and it is now used effectively for some naval and military projects. A full account of the procedure may be found in *Naval Weapons Specification* NWS10, 1971, 3rd edition.

DDS uses a hierarchical approach to the recording of design information. The levels move from the general to the detailed: from the highest and broadest level (showing the system itself) through as many intermediate levels as are considered necessary, to the lowest (illustrating the smallest detail). The format used will depend upon the engineering discipline involved, and whether the subject matter is software, hardware, function or logic orientated. Advantages of DDS include:

- Ease of update.
- Results of studies or analysis may be incorporated.
- Evolves with the project, allowing a dynamic approach to documentation.
- Provides a complete record of the design stages.
- Enables a testing philosophy to be formulated at an early phase of development.
- Allows a wide choice of record format.
- Co-ordinates all design information without necessarily replacing standard publications' procedures.
- Presents maximum information in minimum time.
- Excellent for monitoring progress.
- Encourages a systems approach to design which smoothes over any interface problems.

DDS is not used extensively, but an author will probably come across it at some stage in his career, particularly if he works on MOD documentation.

Diagnostic and maintenance documentation

The job of the maintenance engineer has changed dramatically over recent years. With the increased complexity and turnover of hardware devices, it has not been possible to train him sufficiently within the time available as in more expansive days. One of the consequences of this is that the diagnostic/maintenance engineer now develops a range of 'system skills' at the expense of a thoroughgoing knowledge of discrete component technology. He also makes use of a number of modern aids, including specifically designed documentation.

One of the simplest and most cost-effective ways of maintaining a complex system is by the use of an algorithm* (usually expressed as a flowchart), or a Functionally Identified Maintenance System (FIMS).

An algorithm is a step-by-step procedure, or list of instructions, for a process. A recipe is an algorithm; so is the mathematical base of a computer program.

A flowchart is a diagram representing an algorithm (see *Figures 2.1., 3.1.* and *4.1*). Certain symbols are taken to depict nodal points in the process. A diamond shape signifies a decision point, the question being formulated within the block and two of the angles representing 'yes' and 'no' exits. Similarly, a circle is used for a terminal point, either the beginning or end of a process, or as a link to another sheet. Flowcharts are useful, not only in computer programming, but also as a means of simplifying a complicated diagnostic/maintenance procedure. In such a system, the engineer is given a series of specific actions or observations. Questions are posed, to which the answer is either yes or no – proceed or perform some action.

A related method, but of wider scope and applicability, is the Functionally Identified Maintenance System. The FIMS concept is based on functional flow diagrams which depict *functional sequences* rather than hardware boundaries. The documentation provides the maintainer with a coherent test strategy, logically presented at various levels of complexity. Each level follows on from the preceding one, allowing a structured approach to maintenance not unlike the program steps running a computer.

Fault diagnosis is a logical process beginning with an 'overview', and moving down to the lower, or detailed, levels. Certain definite stages may be isolated:

● Collate* and analyse symptoms.
● Examine equipment.
● Locate fault(s) and cause(s).
● Repair or replace faulty part or unit.
● Test performance.

FIMS documentation reflects this process using a variety of formats at each of three or more levels:

● High or 'master' level – general overview showing major functions.
● Intermediate level(s) – functional subsystems within each major function.
● Low level – the most detailed level.

Functional block diagrams depict functional flows in a left to right direction. Hardware configurations are not relevant to the flow of the diagram, but may be indicated in some way within it.

Functional block texts consist of blocks containing textual descriptions of useful testing information.

Maintenance dependency charts show, in graphical form, the functions and the events which define the dependencies between them. It is a symbolic charting of the functional block diagram.

Symptom analysis charts sometimes replace the maintenance dependency charts at the higher levels.

Test data charts list test points, setting-up procedures, and interpretation of results.

Layout diagrams revert to hardware boundaries enabling the maintainer to locate precise repair points.

Fault-finding and repair are not always carried out by the same technician. Consequently, a division is usually necessary within the documentation to reflect the needs of both diagnosis and maintenance.

Network planning

Whenever a number of people co-operate on a project to a timescale, and with a common end in view, it becomes necessary to introduce some form of planning. Effort needs to be dovetailed, resources allocated with maximum cost-saving efficiency; foreseeable breakdowns avoided, and any frustrations or wastage eliminated.

In industry, the most useful methods of planning involve the construction of a network*. Technical authors may have to work within the compass of a project network, or, if the documentation itself is of sufficient complexity, construct one themselves. This section, then, briefly outlines the salient areas of network planning techniques. More detailed readings can be obtained from the books listed in the Bibliography.

At its broadest, planning usually involves the following elements:

- Objectives.
- Necessary steps to meet objectives.
- Estimates of time and resources needed for each step.
- Risks and contingencies.
- Total time required.
- Total cost.
- Alternatives.
- Decision.
- Establishing schedules.

As a further refinement we may say that a project will be based on:

- Policy.
- Objectives.
- Planning.
- Scheduling.
- Control.

The relationship between these various elements is essentially dynamic and in a state of constant tension. In large projects, imperfections and deviations are always present. Efficient linkage and channels of communication are therefore crucial if the whole edifice is not to crumble into chaos.

One of the methods employed to achieve this is the network plan. As constituted, the planning network forms a subtle and responsive management control system, integrated in depth, and allowing analysis of data and subsequent control of uncertain situations. A carefully constructed network will:

- Define future work.
- Compare supply and demand of resources to improve schedules.
- Improve logistics of resource supply.
- Allow tighter financial control.
- Monitor project progress.

In most forms of networking, nodal points are defined (usually boxes or circles) representing specific events or distinct particulars of a project. Arrowed lines between the boxes signify activities or elements required to attain the nodal points. These relationships are constructed at first without reference to time or dates. Subsequent analysis of events in terms of time reveals a series of concurrent and interconnected processes, each with its own scheduling logic. The total time to complete the project is represented by the most time-consuming path through the network – the 'critical path'*. This is the minimum time necessary to finish the job.

Other, shorter, paths have certain amounts of leeway built into them – known as *slack* or *float* time. Any delay here will not imperil the project's completion date. but, of course, a hold-up on the critical path will inevitably prolong the project and disturb the schedule. This method of networking is known as the Critical Path Method (CPM) or Critical Path Analysis (CPA).

136

Another well-used system is the Project Evaluation and Review Technique, or PERT. The main difference between the two methods is that PERT is *event-orientated* in that it analyses the events or turning-points of a project, while CPA is *activity-orientated*.

Networks may also be hierarchical, or subdivided into levels – as with DDS and FIMS. There are *multi-level* networks, in which the subdivisions are in tiers of higher and lower level networks, a sort of three-dimensional formation.

A further variant is the *sectionalised* network, which simply splits the total plan into small parts for the sake of convenience. An author may well get such a piece of a larger network depicting the documentation schedules and activities.

It sometimes happens that different, though related, projects are linked by consequence of common management or resources, or simultaneous completion dates. In these cases, multi-project networks of great sophistication are often devised. Usually, a computer is brought in to sort out the immensely complicated calculations and consequences. This is achieved by constructing a mathematical model in which every element and problem is logically defined. Every intangible is reduced to numerical data, and all imprecisions quantified. Naturally, errors creep into this process. Ultimately, human instinct and intuition have to be applied to the problems, especially when interpreting results and analyses.

Summary

Translations

Types of translation:

- Literal translation
- Free translation
- Partial translation

Choosing a translator: Translators may be found through

- Local contacts
- Yellow Pages or other directories
- British Standards Institution
- Embassies and foreign legations

Costs of translators: Translations are charged as a rate per thousand words or per folio (75 words). Cost assessment is based on

- Time factors
- Technical content
- Specialised subject matter
- Specifications
- Language
- Other tasks

Indexing

Types of index:

- General index at back of end pages
- Contents list/index

Conventions of indexing:

n – subject included in footnote

ff – used if a number of small inclusions occur over a sequence of pages

passim – used instead of page numbers if the topic is mentioned continuously throughout the book

Development Documentation System

Advantages of DDS:

- Results may be incorporated
- Evolves with the project
- Provides a complete record
- Testing philosophy may be formulated at an early date

- Allows a wide choice of record format
- Co-ordinates all design information
- Presents maximum information in minimum time
- Excellent for monitoring progress
- Encourages a system approach to design

Diagnostic and maintenance documentation

Algorithm: a step by step procedure; a recipe.
Flowchart: a diagram representing an algorithm.
Functionally Identified Maintenance System (FIMS): based on
functional flow diagrams which depict functional
sequences rather than hardware boundaries.
Fault diagnosis: The stages of fault diagnosis are:

- Collate and analyse symptoms
- Examine equipment
- Locate fault(s) and cause(s)
- Repair faulty part
- Confirm performance

FIMS levels:

- High or master level
- Intermediate levels
- Low level

FIMS formats:

- Functional block diagram
- Functional block text
- Maintenance dependency chart
- Symptom analysis chart
- Test data chart
- Layout diagram

Network planning

Basics of a project:

- Policy
- Objectives

139

- Planning
- Scheduling
- Control

Basics of planning:

- Objectives
- Steps to meet objectives
- Estimates of time and resources
- Risks and contingencies
- Time required
- Total cost
- Alternatives
- Decision making
- Schedules

Planning network A network will:

- Define work
- Compare supply with demand
- Improve logistics
- Tighten financial control
- Monitor progress

Critical path analysis: A form of activity-orientated network (CPA or CPM).

Project (or programme) evaluation and review technique: An event-orientated network.

Multi-level network: Subdivided network in the form of tiers or levels.

Sectionalised network: A network which is split into smaller more easily handled parts.

Multi-project networks: Networks cutting across project lines.

8

Technical writing as a career

Modern information technology has not yet affected the traditional white collar professions in the same way as it has the jobs of manual workers. So far, management, and the class who work with the brain rather than the hand, have not suffered the indignity of being replaced by a robotic arm and a microchip.

Technical writing is rather fortunate here; it rides on the back of the new technology. The more technology there is the more work comes to the technical author. Processing machines have yet to reach the areas of creative thought and expression, and, as a guess, it may not be over-optimistic to declare that they never will.

Writing, then, is likely to remain a stable and continuing occupation for some time. New recruits may hope to avoid the Luddite sense of insecurity affecting other trades and professions. The future is, at least, hopeful, because of the direct connection between technical change and the quantity of writing work available at any one time. No profession, however, is immune from the passage of time and the inevitability of change.

Two distinct trends can be detected in the area of writing for modern technology. On the one hand, there is a lobby who believe that a shortened technical language and syntax holds out the best hope for a precise science of technical description. On the other, developments in theoretical physics point to a more relativistic, even metaphorical, approach to our explanation and understanding of the world.

Software is being asked increasingly to imitate the infinite possibilities of reality instead of the straight-laced linear concepts of traditional engineering.

Hardware, albeit Lego-like in aspect, is now rooted firmly in the sub-atomic environment, where the concepts of classical mechanics lose ground before a wild dance of unfamiliar energies and forces.

Figure 8.1. Modern technology in authorship (by courtesy of Hewlett-Packard Ltd)

Heisenberg summarised the paradox thus: 'The problems of language here are really serious. We wish to speak in some way about the structure of atoms . . . but we cannot speak about atoms in ordinary language.'

Fritjof Capra, in his book *The Tao of Physics*, explains it as follows: 'The notion that all scientific models and theories are approximate and that their verbal intepretations always suffer from the inaccuracy of our language was already commonly accepted by scientists at the beginning of this century, when a new and completely unexpected development took place. The study of the world of atoms forced physicists to realise that our common language is not only inaccurate, but totally inadequate to describe the atomic and sub-atomic reality. Quantum theory and relativity theory, the two bases of modern physics, have made it clear that this reality transcends classical logic and that we cannot talk about it in ordinary language.'

That this theoretical sophistication will eventually filter through to the down-to-earth concepts of engineering is certain, and probably only a matter of time. The technical author may then be given the task of trying to explain the unexplainable; of developing metaphors and analogues for deeply paradoxical situations.

As science reassesses itself in terms of its wider implications, engineering too will need to adapt to new complexities and possibilities. Language will be forced to expand and develop in an attempt to remain abreast of the new philosophies. Even simple technical descriptions, essential for complete understanding, will need to reflect the depth of theoretical complexity inherent in modern devices.

In future, much of a technical writer's work will probably be done at the keyboard of a word processor. The devices described will all partake of the *same* technology. It follows that a basic knowledge of microprocessor techniques will be a standard requirement for the job. And since software is a form of deferred design, it is likely that hard and soft will not retain the distinctions they now enjoy.

The technical author of the near future will be a resourceful individual. He will need to span disciplines even more than he does now. But while techniques move on, and change is there to be described, he will survive. And, with a little bit of luck, prosper.

9

Useful information

Courses and examinations

The City and Guilds of London Institute runs two examinations for students wishing to qualify in Technical and Scientific Communication. The examinations are:

- Technical communication techniques – 536–1.
- Technical authorship – 536–2.

Courses specifically designed for these qualifications are held at the following technical colleges:

Bradford College of Art and Technology (536–1)
Chippenham Technical College (536–1)
Corby Technical College (536–1)
Farnborough College of Technology (536–1)
Glasgow College of Building and Printing (536–1)
Grantham College of Further Education (536–2)
Highbury Technical College, Portsmouth (536–1&2)
Hull College of Further Education (536–1)
Kitson College of Technology, Leeds (536–1)
Norfolk College of Art & Technology, King's Lynn (536 –1&2)
Norwich City College (536–1&2)
Openshaw Technical College, Manchester (536–1&2)
Stevenson College of Further Education (536–1)
Willesden College of Technology (536–1&2)

Self-contained courses which are recognised by the City and Guilds Institute as qualifying for exemption from part of their examination, are run by:

Intereurope Technical Services,
High Walls,
East Street,
Fareham,
Hants, PO16 OBZ.
Tel: Fareham 232336.

Module 1: Scientific and Technical Communication.

'This six-week full-time course, available for sponsorship under the government's Training Opportunity Scheme (TOPS), academically prepares engineers from industry and the Services for a career in the Technical Publications Industry.'

Module 2: Digital Techniques and Introduction to Computers.

'This two-week full-time course provides theoretical tuition (backed by practical proving demonstration) to furnish students with a basic knowledge of digital techniques and the more common types of logic elements.'

The two modules are usually taken in one eight-week stretch, with the final fortnight serving to update the student's technical know-how.

The City and Guilds Institute describe the courses in Technical Authorship as follows:

'Candidates for the examination in Technical Authorship will normally be aspiring or practising technical authors. As such, they will possess technical qualifications and experience befitting their particular technologies. The aim of the course is to provide instruction and qualification in technical authorship per se although it is realised that many of the examples and exercises will necessarily be taken from a technological environment.'

The breakdown of each course is as follows:

Technical communication techniques –

- Recognition of the problems of communication
- Material for communication
- Media for conveying a communication

145

Technical authorship –
- Job definition
- Products of a technical author
- Planning an assignment
- Information gathering
- Style of writing
- Graphical presentation
- Graphical techniques
- Preparation of drafts
- Editing
- Reproduction and finishing processes

Further information on these examinations can be obtained from:

The City and Guilds of London Institute,
76 Portland Place,
LONDON, W1N 4AA.

There follows a selection of questions from past papers of both examination subjects:

Technical communication techniques –

- List six points which must be taken into consideration in the editorial checking of technical documents.

  (3 marks)

- Briefly define THREE of the following
  (a) collation
  (b) pagination
  (c) loose leaf
  (d) recto/verso
  (e) watermark   (6 marks)

- Mention techniques in note-taking that ease assimilation and subsequent memorisation of information.

  (6 marks)

- Itemise FIVE ways of overcoming resistance to the imparting of technical information.   (5 marks)

- List THREE mechanical and THREE other sources from which information may be gathered.   (3 marks)

Technical authorship –

- Explain what you mean by the word 'jargon'. Discuss the use of jargon by scientists and technologists amongst themselves and in communicating with others. Comment on your role as a technical author in dealing with jargon. (25 marks)

- Your company is considering the manufacture of a new product. The basic design has already been formulated and a preliminary meeting has been arranged to discuss development and production. Your task will be to prepare installation, operating and maintenance documentation for the product and you are to attend the preliminary meeting.

  Prepare a set of notes to serve as a basis for discussion at the meeting, outlining the type of information and facilities which you will reasonably expect to be given and any others which you consider will help in the preparation of the documentation.

  (25 marks)

- A technical author may sometimes be called upon to appraise and update a handbook prepared some time previously by another author.

  (a) Prepare a set of simple recommendations for an author undertaking such a task for the first time.

  (b) Indicate how the author should proceed with the task in an effective and constructive way. (25 marks)

The foregoing extracts from City and Guilds documentation are reproduced by kind permission of the Institute. All enquiries should be addressed to the Institute at the address given above.

Professional body

The Institute of Scientific and Technical Communicators (ISTC) is the professional body representing many technical authors and communicators. It aims to establish ethical and professional codes of practice within the industry. It is part of the international body INTECOM which performs similar functions overseas in such countries as the USA, Canada, France, Germany, Holland, Denmark, Norway, Sweden and Australia. Further information may be obtained from:

The Secretary,
Institute of Scientific and Technical Communicators,
17 Bluebridge Avenue,
Brookmans Park,
Hatfield,
Herts, AL9 7RY.

Useful addresses

British Standards Institution,
2 Park Street,
London, W1A 2BS.

International ISBN–Agentur,
Staatbibliothek Preussicher Kulturbesitz,
1 Berlin 30,
Potsdamerstrasse 33,
West Germany.

Standard Book Numbering Agency Ltd.,
12 Dyott Street,
London, WC1A 1DF.

Translators' Association,
84 Drayton Gardens,
London, SW10 9SD.

Translators' Guild Ltd.,
24A Highbury Grove,
London, N5 2EA.

Writers' Guild,
430 Edgware Road,
London, W2 1EH.

Society of Authors,
84 Drayton Gardens,
London, SW10 9SD.

Society of Indexers,
The Registrar,
25 Leybourne Park,
Kew Gardens,
Surrey, TW9 3HB.

Patent Office,
(Dept of Trade),
25 Southampton Buildings,
London, WC2A 1AY.

Dept of Education and Science,
Elizabeth House,
York Road,
London, SE1 7PH.

Ministry of Defence (Press & PR)
(General, Navy, Army & Air Force),
Main Building,
Whitehall, London, SW1A 2HB.

Copyright Receipt Office,
The British Library,
2 Sheraton Street,
London, W1V 4BH.

British Library Lending Division,
Boston Spa,
Wetherby,
W. Yorks, LS23 7BQ.

Science Museum,
South Kensington,
London, SW7 2DD.

National Library of Scotland,
George IV Bridge,
Edinburgh, EH1 1EW.

Science and Engineering Research Council,
Polaris House,
North Star Avenue,
Swindon, SN2 1ET.

HM Stationery Office,
Sovereign House,
Botolph Street,
Norwich,
NR3 1DN.

Glossary

Abstract Abbreviated form of document. A summary or digest with reference to sources.

Addendum (Pl. Addenda) Additional material; update or afterthought.

Algorithm Step by step procedure or process. Used as the basis of flowcharts and computer programs.

Ampersand The shape of the character '&' ('and per se – and' or '& by itself = and').

Annex(e) A supplementary section. Used in the Ministry of Defence JSP specification system.

Annotation A note attached to an illustration for purposes of explanation or comment. Direct ––: a non-keyed annotation. Indirect ––: a keyed annotation.

Appendix (Pl. Appendices) An addition to the main body of a document, having some contributory value, but being tangential to the broad argument of the work. Often contains detailed information for the sake of completeness or expansion.

Article A short piece of writing with a single theme, usually written for a magazine or newspaper.

Aspect ratio The height/width ratio of an illustration.

Atlas Abbreviated Test Language for All Systems. An MOD approved shortened vocabulary and syntax system.

Audio-visual A form of presentation in which slides or other visual material are linked to a spoken commentary.

Author One who sets forth written statements; the writer or composer of a treatise or book.

Author's corrections Corrections introduced at printer's proof stage derived from the author's mistakes or omissions.

AvP 70 Specification for military aviation manuals.

Backed up A sheet printed on both sides.

Bank paper Thin typing paper used for carbon copies.

Bibliography Collection of book titles and other information. Usually thematically related, or relevant to the contents of a book.

Bind The process of fastening pages together inside a form of cover.

Body matter The textual content of a page excluding titles, headings and so on.

Bold typeface The heavier version of a typeface.

Bond paper Top copy paper of medium weight.

Broadside Large paper size from which other sizes are derived by progressive division. Used in the British system of paper sizes.

Bromide A monochrome photographic print.

BS 4884 British Standard for technical manuals and handbooks.

Bug Error in computer program.

Camera copy Sometimes called Camera-Ready Copy. Copy suitable for reprographic or printing processes.

Caption A heading of some kind, or a description of an illustration.

Cast off An estimate of the length of a piece of writing based on a careful review of the text.

Catalogue A systematic list of a collection of documents.

Centre heading A title or heading centrally placed on a page or column.

Clean proof A correct proof copy needing no alteration.

Collate An ordering process carried out on a batch of pages to form copies of a book or section.

Colour separation A set of drawings of different colours used for the printing of an illustration.

Combination line and tone An illustration which uses both line and half-tone.

Condensed   A form of letter style in which each character is reduced.

Continuous tone   Complete gradation of tone in an illustration.

Copy   Text or matter intended for printing.

Copyright   Legal prerogative granted to the author of a work.

Copywriting   An expressive form of writing, usually sales orientated, used for glossy brochures, etc.

Critical path analysis   An activity-orientated network.

Cut away   A form of illustration in which the outer casing of the depicted equipment is cut away to reveal the internal components.

Diagram   Two-dimensional symbolic representation.

Documentation   Collection of documents on a given subject. Also used to describe the organisation, recording and generation of documents for the purpose of storage, retrieval, use or transmission of information.

Double page spread   An illustration running across two pages.

Draft   An early version of a document drawn up for initial consideration.

Dummy   Or Printer's Dummy. A complete sample mock-up of a proposed book.

Dyeline   Or Diazo process. A copying process for engineering drawings.

Editing   The revision and alteration of a draft document prior to publication.

Enlargement ratio   The ratio of enlargement in linear dimensions of an illustration or copy.

Errata   A list of corrections to be made in a text. Usually attached to a book in the form of a pasted on sheet.

Expanded   A wider version of a type style.

Exploded view   An illustration showing a stage in the dismantling process of a piece of equipment.

Filmsetting   Method of typesetting on film or paper prior to platemaking.

Floppy disc A recording medium consisting of a pliable disc coated with ferrous oxide and used to store information magnetically.

Flowchart The graphic representation of an algorithm.

Folio 1. A manuscript page number. 2. Unit of manuscript length (75 words). 3. A sheet of paper folded once.

Font A set of type characters in one size and style.

Foolscap A paper size (13 × 8 inches); no longer used.

Footnote A note or reference placed at the foot of a page.

Foreword An introductory section to a book, usually written by someone other than the author.

Format Layout, shape and size of a publication.

Fount See *Font*

Frontispiece A general illustration placed at the front of a book, and not part of the text.

Galley proofs A copy of printed material taken when the matter is set up in the galley.

Glossy A high cost publication, usually a brochure or sales document, remarkable for its glossy finish.

Graphics The artwork contained in a publication.

Graticule The grid on a graph or map.

Gravure An intaglio printing process in which the image is engraved on a plate.

Hard copy Used to describe a printed sheet as opposed to a soft copy on a VDU.

Hardware The physical equipment of any system.

Half plate Photographic material measuring 6½ × 4 inches.

Half title A recto page in a book bearing the title only.

Half-tone A continuously toned print in which the tonal gradations are represented by dots of varying size.

Heading weights The importance of a title expressed in terms of its heading status or prominence.

Imprint The statement of the name of the publisher, the place of publication, and the date appearing in a book.

Indentation The beginning of a line a number of spaces inset to the right to indicate a new paragraph or emphasise the contents.

Index A list of subjects together with their page references.

Inking The process of applying ink to a printing machine.

Intaglio A printing process in which the image is engraved or etched below the surface of the plate or block.

Issue A stage in the life of a publication.

Italic A typeface which slopes to the right. Used to single out parts of the text.

JSP Joint Services Publication – specifications of the Ministry of Defence.

Justification A method of composing to align each line of print immediately below the one above. Right-hand justification assumes that each line is of precisely the same length.

Key Indirect annotation.

Landscape An illustration with its longer edge horizontally arranged.

Laymark A frame or mark in a non-reproducible colour printed on a camera copy sheet.

Layout The arrangement of illustrations and text.

Leader line A line in an annotated illustration pointing to a specific item.

Letterpress A method of printing in which the letter or character is raised in relief above the surface of the type.

Line illustration An illustration consisting entirely of line and possessing no tonal gradations.

Line work The artwork associated with line illustrations.

Lithography A planographic printing process in which the image is formed on the surface of the plate.

Loose leaf A binding method which allows pages of a book to be removed or added.

Lower case Non-capital letters.

Machine proof A proof taken on the production printing press to indicate the quality of the finished product.

Mark-up Instructions written on a document at proof stage intended for a typist or printer.

Master Original camera copy from which further copies can be made if required.

Mechanical tint Self-adhesive transparent sheet with patterns of dots or line which can be transferred by pressure onto an illustration.

Microfiche A micro-recording method consisting of a flat, transparent sheet of film onto which data has been photographically impressed.

Microform The method of storing information on photographic film.

Mock-up A specimen layout of a document.

Modification A change made to a piece of equipment which usually requires some change to be made in the documentation.

Montage A fitting together of different images to make a whole.

Negative A reversed photographic image.

Network planning A method of recording information and decision taking based on the drawing of a network. CPA and PERT are most widely used systems.

Octavo (8vo) A paper size that is ⅛ of a Broadside.

Offset litho A lithographic printing process in which the ink is transferred to an intermediate roller before application to the paper.

Original A document to be copied or printed.

Orthographic A one-view method of illustration. Other views are shown separately.

Overlay A transparent sheet placed over an illustration for annotation purposes, or adding further information or colours.

Page proof A proof stage in the printing of a publication when the copy is set up into pages.

Pagination The method of page numbering.

Paste up A layout composed of pasted on items indicating the final look of a printed document.

PERT Project Evaluation and Review Technique.

Perspective   The representation of three-dimensional space on a two-dimensional surface.

Photogravure   A method of printing in which the image is applied to a plate photographically before engraving.

Pica   A measurement of type composition (12 points or ⅙ of an inch).

Point size   Unit of type measurement (¹⁄₇₂ of an inch).

Portrait   A page arrangement in which the shorter edge is horizontal.

Preface   An introductory section to a book, usually stating the purpose or origin of the work.

Prelims   The preliminary pages of a publication.

Presentation   1. The style and format of a book. 2. A lecture or talk in which certain information is disseminated.

Program   A set of steps and procedures running a computer.

Proof   A sample copy taken prior to printing.

Proportional spacing   A method of arranging type such that each character takes up a width in proportion to its actual dimensions.

Quarto   A paper size made by folding a Broadside sheet into four.

Ream   500 sheets of identical paper.

Recto   The right hand page of an open book.

Reduction   The process of making a smaller image from copy or illustrations.

Register   Correct alignment of two images printed on the same sheet.

Relief   Printing processes in which raised surfaces are inked before application to the paper.

Reprography   Copying processes using light or other forms of radiation, such as photocopying. Also includes office duplicators and copiers.

Rough (visual)   An initial sketch allowing an assessment to be made of artwork.

Running head   A heading which runs throughout the pages of a book, such as a company logo.

Run on   Continue: either typing or printing.

Saddle stitching A method of joining a single section book by stapling through the fold.

Secondary master A copy taken from the master for the purpose of amending and taking further copies.

Software All the components of a computer system not defined as 'hardware'. Includes programs, compilers, and documentation.

Specification A set of requirements outlining and describing the methods to be used in the production of a publication.

Subtitle A secondary title expanding the main title.

Synopsis A precise summary of the contents of a book.

Throw clear A folded illustration which may be extended such that the whole of it is visible when the book is closed.

Throw out An illustration which may be extended beyond the width of the book, but which is printed up to the binding edge.

Title page The sheet at the beginning of a document which bears the main title.

Twice up An enlargement size for drawings which are then reduced in the printing process.

Typeface A set of characters of a particular design available in varying sizes.

Typography The process of selecting type for printing work.

Upper case Capital letters.

Verso The left hand page of an open book.

Vetting The process of validating a text.

Wash A tone applied by brush or airbrush during artwork.

Watermark A distinct mark visible against the light in the fabric of typing or printing paper.

Word processor A computer interfaced to a keyboard and printer, and containing a sophisticated editing program.

Writer One who sets forth written statements, the author or composer of a treatise or book.

Bibliography

This list is a select bibliography ranging over most of the topics covered in this book. It is by no means exhaustive, and public library catalogues should be consulted for a wider survey of the field. The books listed here are generally well regarded by practitioners and should prove useful for further study.

Aslib Directory, Volume 1, Science, Technology and Commerce; 4th ed. 1977.

Baker, C. *A Guide to Technical Writing*; Pitman.

Baker, C. *Technical Publications*; Chapman & Hall.

Banks, J. G. *Persuasive Technical Writing*; Pergamon Press.

Beck, G. J. N. *Preparation of Scientific Material for the Press*; ASLIB.

Bowman, W. J. *Graphic Communication*; Wiley, 1968.

Carey, G. V. *Making an Index*; Cambridge University Press.

Cooper, B. M. *Writing Technical Reports*; Pelican, 1964.

Darbyshire, A. E. *Report Writing – The Form and Style of Efficient Communication*; Edward Arnold, 1970.

Ewart, K. *Copyright*; Cambridge University Press, 1952.

Flint, M. F. *A User's Guide to Copyright*; Butterworth, 1979.

Glidden, H. K. *Reports, Technical Writing, Specifications*; McGraw Hill, 1964.

Hart, Horace. *Rules for Compositors and Readers at the University Press*; 38th ed. Oxford University Press, 1978.

Henn, T. R. *Science in Writing*; Harrap, 1960.

Hoffmann, Ann. *Research; A Handbook for Writers and Journalists*; 2nd ed. Black, 1979.

Hodgson, Francis. *Industrial Research in Britain*; 8th ed. 1976.

Holmes, C. *Beginner's Guide to Technical Illustration*; Newnes Technical Books, 1981.

Kapp, R. O. *Presentation of Technical Information*; Constable.

Klare, G. R. *The Measurement of Readability*; Iowa State University Press, 1963.

Leyton, A. C. *The Art of Communication*; Pitman, 1968.

Lockyer, K. G. *An Introduction to Critical Path Analysis*; Pitman, 1964.

Lucas, F. L. *Style*; Cassell.

Oxford Dictionary for Writers and Editors; OUP, 1981.

Quiller-Couch, A. *On the Art of Writing*; Cambridge University Press, 1916.

Scientific and Technical Communication; Intereurope Technical Services, 1980.

Sklare, A. B. *Creative Report Writing*; McGraw-Hill.

Publishing Contracts; Society of Authors, 1974.

Standards for Authors and Printers; H.M.S.O.

Thomas, David St John. *Non-Fiction – A Guide to Writing and Publishing*; David & Charles, 1970.

Unwin, Philip. *Book Publishing as a Career*; Hamish Hamilton, 1965.

Unwin, Stanley. *The Truth about Publishing*; Allen & Unwin.

Whitehouse, F. E. *Documentation*; Business Books, 1971.

Williams, G. E. *Technical Literature – Its Preparation & Presentation*; Allen & Unwin.

Williams, P. T. *The Practical Technical Author*; TASS (AUEW) 1981.

Woodgate, H. S. *Planning by Network*; 2nd ed. Business Books, 1967.

Index